Cosmology and the Scientific Self in the Nineteenth Century

Howard Carlton

Cosmology and the Scientific Self in the Nineteenth Century

Astronomic Emotions

palgrave
macmillan

Howard Carlton
Kidderminster, Worcestershire, UK

ISBN 978-3-031-05279-8 ISBN 978-3-031-05280-4 (eBook)
https://doi.org/10.1007/978-3-031-05280-4

Cover illustration: Marina Lohrbach_shutterstock.com

This Palgrave Macmillan imprint is published by the registered company Springer Nature Switzerland AG.
The registered company address is: Gewerbestrasse 11, 6330 Cham, Switzerland

Acknowledgements

I would like to record my gratitude to the people and institutions who provided generous advice and assistance during the development of this book.

Firstly, I must thank Dr Rob Boddice, Dr David Gange and Dr Sadiah Qureshi, for their constant encouragement and their very welcome suggestions, corrections and thought-provoking interventions which have guided me through what was a lengthy but enjoyable gestation.

My thanks are also due to several people who have provided informed and detailed commentary, via email in some instances, or perhaps over a cup of coffee or glass of wine during a conference: Dr Simon Briercliffe, Dr Nicholas Campion, Professor Michael J Crowe, Professor Gowan Dawson, Dr Eleanor Dobson, Dr Rhodri Hayward, Dr Michael Hoskin, Dr Ruth Lindley, Dr Lee MacDonald, Professor Matt Houlbrook, Professor Rosemary Mitchell, Dr Joshua Nall, Dr James Pugh, Professor Crosbie Smith, Professor Robert Smith and Professor Bennett Zon.

I also want to acknowledge the assistance of the staff of the following institutions who facilitated the use of their respective archives:

- Brotherton Library Special Collections, Leeds University—Clodd correspondence
- The Herschel correspondence in the library of The Royal Society.
- The archives of Birr Castle, Birr, Co. Offaly, Republic of Ireland
- The Syndics of Cambridge University Library.

- The University of St Andrews Libraries and Museums
- Excerpts from the correspondence between John Frederick William Herschel and John William Whittaker are reproduced by permission of the master and fellows of St John's College, Cambridge

A paper based on elements of Part III, specifically the narrative of Nichol's response to the claimed refutation of the nebular hypothesis, has previously been published as: Howard Carlton, 'Condensing from a Fluid Haze: John Pringle Nichol, the Nebular Hypothesis and Nineteenth-Century Cosmogony' in Nicholas Campion and Chris Impey (eds.), *Imagining Other Worlds: Explorations in Astronomy and Culture* (Ceredigion: Sophia Centre Press, 2018), 87–94.

Finally, I would like to record my special thanks to friends and family for their encouragement and interest and in particular to my wife Penny who has supported me in every way throughout the project.

About the Book

This book will argue that while the historiography of the development of scientific ideas in the nineteenth century has for some time acknowledged the important influences of social, cultural and material factors, and the recent turn to the history of experience has further reinforced the influences of internal and external psychotropic stimuli on subjectivities, the significant impact of traumatic events, life-threatening illnesses or bereavement on the development of scientific thought may not have been fully recognised. The ideas of selected subjects with respect to the nature of the universe and the interlinked epistemologies of science, religion, metaphysics and political ideology were sometimes materially altered by affective life-events as evidenced by the positions which they subsequently adopted and promulgated in relation to these overlapping sources of knowledge and praxis.

This work will focus in particular on the influence of such events on the developing views of several nineteenth-century scientists and science writers who opined on the structure, size and composition of the universe. The approach taken will be to explore three topics of great interest to astronomers and physicists during this time period: the question of the possible existence of life on other planets, the deployment of the nebular hypothesis as a theory of cosmogony and the religiously charged debates about the ages of the earth and sun.

Evidence has arisen as a result of a close reading of the comments and narratives of the subjects and their biographers for the sometimes-subtle and complex influences which influenced their thinking on these questions. In the course of studying such primary sources, it became clear that they frequently referred to significant life crises, sometimes accompanied by the consumption of psychoactive substances, which preceded important changes in their theological, metaphysical and scientific worldviews. A few notable cases of this ilk have already been signposted by historians, but the case histories discussed in this book will provide a significant set of new examples of such affects and effects. We will uncover the stories of several such perspective-altering crises and will further interrogate the available documents in order to study the relationships between mind and body thus revealed. Applying this suggested mode of interpretation to the broader field of nineteenth-century scientific developments in general, we shall achieve a greater understanding of the underlying forces which drove the intellectual history of that era and which are still relevant to today's knowledge-making processes.

Contents

PART I

Introduction

CHAPTER 1

Crisis and Cosmology: The Subtle Interactions of Mind, Body and Belief

In the last short illness, her conduct in simple truth was angelic. She never once complained; never became fretful; was ever considerate of others, and was thankful in the most gentle, pathetic manner for everything done for her. When so exhausted that she could hardly speak, she praised everything that was given her, and said some tea 'was beautifully good.' When I gave her some water, she said, 'I quite thank you;' and these, I believe, were the last precious words ever addressed by her dear lips to me.
Charles Darwin, *Life and Letters*[1]

On the 30th of April 1851, Charles Robert Darwin (1809–1882) wrote a "short sketch" to capture the feelings of "very severe grief" he was experiencing following the recent death of his daughter as "she finally succumbed to tuberculosis." He had closely observed Annie (1841–1851) during the last few feverish weeks of her life, and he could only confide his anxiety to his pregnant wife Emma (née Wedgwood, 1808–1896), who had not accompanied him to Malvern, via a series of "poignant letters". His memoir was "a stunning vindication" in which "Annie appears as a type-specimen of all the highest and best in human nature."[2]

[1] Francis Darwin, (ed.), *The life and letters of Charles Darwin, including an autobiographical chapter*, vol 1 (London: John Murray, 1887), 134.

[2] Nora Barlow, *The autobiography of Charles Darwin 1809–1882. With the original omissions restored. Edited and with appendix and notes by his grand-daughter, Nora Barlow.* (London: Collins, 1958), 97–98. John H. Brooke, 'Darwin and Religion: Correcting the Caricatures', *Science & Education* (2010), 19: 391–405 at 397. https://doi.org/10.1007/

H. Carlton, *Cosmology and the Scientific Self in the Nineteenth Century*, https://doi.org/10.1007/978-3-031-05280-4_1

Although it has always been evident from both contemporary accounts and subsequent scholarship that during his career Darwin suffered a significant lessening of belief in revealed religion, for a long time it had been assumed, based on comments made by Darwin himself, that the process of unconversion occurred relatively early in his career.[3] In an important intervention in 1989, however, James Moore argued that the timing and circumstances of Darwin's final loss of Christian faith needed to be significantly reconsidered.[4] After examining the context and sequence in which Darwin's thoughts on religion were presented in his *Autobiography*, which had not been available in unexpurgated form until 1958, Moore placed the key moments in his changing relationship with religion in a critical period when he was affected by "anger and grief":

> On the one hand, he had drawn the line in the first part of the section on religious belief by angrily rejecting Christianity for its 'damnable doctrine', that would put his father in hell. Dr Darwin [his father] died on 13 November 1848. [...] Annie died on 23 April 1851. Two strong emotions, anger and grief, in the *Autobiography* mark off the years from 1848 to 1851 as the period when Darwin finally renounced his faith. Two deaths, and two births, lie at its extremes.[5]

Annie's death can therefore be seen as the one of the most important steps in a sequence of events which precipitated Darwin's terminal break with the concept of a caring and paternalist Trinitarian god. This break led in turn, Moore argued, to an increased engagement with the secular processes of biological evolution. It has subsequently become almost a historiographical given that Annie's death was the final straw which confirmed Darwin's loss of faith in a Christian theodicy.

The major argument to be presented in this book is that some significant steps in our changing understanding of the universe in the nineteenth century were, in a manner analogous to the narrative of Darwin and Annie, contingent on preceding instances of psychotropic affect including

s11191-009-9211-9. James R. Moore 'Of Love and Death: Why Darwin "gave up Christianity"' in James R. Moore (ed.), *History, Humanity and Evolution: Essays for John C. Greene* (Cambridge University Press, 1989), 195–229 at 218.

[3] Moore, 'Death', 197.

[4] Moore, 'Death', This argument was further developed in Adrian Desmond and James R. Moore, *Darwin* (London: Michael Joseph, 1991).

[5] Moore, 'Death', 209. The passage which Moore used as his source for this reading is: Barlow, *autobiography*, 85–98.

serious illness, injury or anticipated bereavement, sometimes accompanied by the consumption of psychogenic substances. Such changes were generally associated with a notable shift in personal theology which in due course, sometimes after an interval of several years, led to contingent changes in related and intersecting epistemologies including astronomy and the other physical sciences. It should of course be acknowledged that recent historiography has already pointed out the close connections between religious beliefs, political aims and cultural contexts which have influenced various subjects' scientific productions. What does not seem to have been as strongly noted to date, however, is the extent to which significant life events have in due course led to marked changes of previously espoused positions in one or more components of an individual's construction of the cosmos.[6]

Most of us sense that there is more than a passing connection between emotion and cognition. We are also aware that our responses to the external world in which we are situated are driven, at least in part, by our conscious and unconscious interactions with the complex internal systems on which we are dependent for our survival. The SARS-CoV-2 (more generally known as COVID-19) pandemic, for example, has provided ample evidence that infection with a virus can lead to significant and enduring complications which include depression, anxiety and sleep disturbance.[7] In other words, mind and body are not independent dualist entities but are in fact closely bound together. Whilst this point has long been recognised by scholars representing a wide variety of academic fields such as neurophysiology, the social sciences, history and, more recently, immunology, it has not been as clearly recognised that amongst the longer-term results of a variety of adverse psychological and physiological stimuli are significant shifts in such apparently anodyne questions as the nature and origin of the universe.

In order to sustain this line of argument, we will not only examine the histories of a number of subjects whose narrative arcs will, we argue, demonstrate the claimed connection between affect and astronomy, but we will also look at some of the underlying causes which have explanatory

[6] It is, however, worth noting that similar arguments have been and can be made for the influence of affective stimuli on political thinking. See: Ruth Leys, *The Ascent of Affect* (University of Chicago Press, 2017), 309.

[7] Jiawen Deng et al., 'The prevalence of depression, anxiety, and sleep disturbances in COVID-19 patients: A meta-analysis', *Ann. N.Y. Acad. Sci.* xxxx (2020) 1–22. https://doi.org/10.1111/nyas.14506.

power when we endeavour to unravel such complex linkages. To begin with, it will be important, going forward, to establish some theoretical understanding of how a nineteenth-century person when faced by an existential challenge might not only modify their personal religious stance but, rather less predictably, also go on to alter their ontological beliefs.

Experience, Emotions and *Weltmodelle*

A wide variety of terms have been used to describe the response of the brain–body system to affective internal and external stimuli. In the late eighteenth century, "relevant mental states were categorised variously as 'appetites,' 'passions,' 'affections,' or 'sentiments'". During the nineteenth century, the term 'emotions' began to be used, firstly by psychologists and then by historians, as a portmanteau term to cover approximately, but not precisely, the same gamut of meaning as the preceding terms.[8] More recently, there has been a move to combine the ideas associated with the history of emotions with related concepts, such as the history of the senses, into a broader based history of experience.[9]

The physical underpinnings of our perceived experiences may potentially be found in the hypothesis that our short-term reactions to internal and external stimuli are anticipated and predicted on the basis of our previous exposure to similar situations. This prediction process is used to anticipate future physiological needs and thus to mediate the combination of affect and cognition which takes place in our 'biocultural' brain–body systems.[10] Many such incidents are tagged with emotion words which relate to preceding occurrences of the same affect.[11] The mind–body system then generates a retrospective narrative of the way in which events appeared to unfold: "to infer the meaning of the sense data and guess

[8] Thomas Dixon, '"Emotion": The History of a Keyword in Crisis', *Emotion Review*, 4, 4 (October 2012): 338–344.

[9] Rob Boddice and Mark Smith, *Emotion, Sense, Experience* (Cambridge University Press, 2020).

[10] Lisa Feldman Barrett, *How Emotions Are Made: The Secret Life of the Brain* (London: Pan Macmillan, 2018). Lisa Feldman Barrett, 'The theory of constructed emotion: An active inference account of interoception and categorization', *Social Cognitive and Affective* Neuroscience, 2017, 1–23. Anil K. Seth, 'Interoceptive inference, emotion, and the embodied self.', *Trends in Cognitive Sciences*, 17, 11 (2013): 565–573.

[11] Katie Hoemann, Madeleine Devlin and Lisa Feldman Barrett, 'Comment: Emotions Are Abstract, Conceptual Categories That Are Learned by a Predicting Brain', *Emotion Review*, 12, 4 (October 2020): 253–255, quoted by Boddice and Smith, *Experience*, 34.

what to do about it".[12] This post facto processing can also supply a veneer of rationality and morality for our actions which overlays our private intentions, regardless of whether or not those intentions surfaced prior to or after the experience in question.[13]

Cognitive processes therefore do not represent some kind of independent, higher-order rationality but are instead intimately bound to the physical and temporal environment of the thinker, resulting in the construction of a reflexive and plastic understanding of the working of the world. For convenience, we shall label this holistic picture with the signifier *Weltmodell* (world model), a German word which has previously been employed to describe both the cosmological models of physical scientists and the creation stories found in traditional religions. It is thus traditionally related to the two key epistemologies under discussion here. We, however, will be using this term to represent a more affective internal representation of external reality than the alternative *Weltanschauung* (worldview), which might be understood to encompass only apparently objective and rational perspectives. In other words, a *Weltanschauung* can be defined and shared verbally, whereas a *Weltmodell* is internally felt or experienced.

The sources which inform a subject's *Weltmodell* are many and varied, but some common threads in the process of formation can be identified. Historians of emotion, including William Reddy, Barbara Rosenwein, Jan Plamper and Rob Boddice have pointed out that there are hermeneutic feedback loops involved in the engagement of subjects with their social networks which lead to them evolving a repertoire of selves with which to respond to the external environment and a shared vocabulary with which to describe their feelings.[14]

[12] Lisa Feldman Barrett, *Seven and a Half Lessons about the Brain* (London: Picador, 2021), 66.

[13] Otniel Dror, 'Historians in the Emotion Laboratory', *Emotion Review*, 12 (2020): 191. Michael Roper, 'Splitting in unsent letters: Writing as a social practice and a psychological activity', *Social History*, 26, 3, October 2001: 318–339, at. 319, 323–324. On the topic of intentionality and anti-intentionality, see, for example, Leys, *Affect*, 335–340.

[14] A brief introduction to this subject can be found in: Jan Plamper, 'The History of Emotions: An Interview with William Reddy, Barbara Rosenwein, and Peter Stearns', *History and Theory*, 49 (May 2010): 237–265. Lengthier texts which provide a general survey of this field include: William Reddy, *The Navigation of Feeling: A Framework for the History of Emotions* (Cambridge University Press, 2001), Rob Boddice, *The History of Emotions* (Manchester University Press, 2018) and Barbara H. Rosenwein and Riccardo Cristiani, *What is the History of Emotions?* (Cambridge: Polity, 2018).

Our *Weltmodelle* also interact with a variety of external sources of apparently stable, but in practice malleable, information about the environment in which we find ourselves. In Thomas Dixon's comprehensive history of the study of emotions in the nineteenth century, he noted that there are three important components of an individual's construction of the world:

> assumptions about the nature of reality (ontological commitments), assumptions about how best to acquire knowledge about reality (epistemological commitments) and commitments to particular textual and metaphorical resources as the best way to articulate knowledge of reality (narrative and explanatory commitments).[15]

The selection of sources to be employed and selves to be performed can clearly be impacted by a significant life event, and it is also evident that for most of our subjects a change in one axis of Dixon's triad sooner or later produced consequent changes in the other domains. One potential result of exposure to the potent affect of a brush with mortality, for example, was the emergence of a changed ontological understanding of the universe wherein the subject's heightened concerns were assuaged by the imagining of the kind of divinely ordered cosmos capable of sustaining eternal human souls which could be found in contemporary and preceding religious tracts. Alternatively, a near-death experience or an anticipated bereavement sometimes leads to increased doubt and reduced religiosity. As we have seen in the case of Darwin, the affect associated with the death of "the joy of the household" suggested to the subject a modified theodicy in which an already-waning belief in a paternalistic, sometimes punitive, deity was significantly reduced, or removed altogether, and the ensuing *Weltmodell* shift prompted an increased reliance on naturalistic ontologies.[16]

Thus, historical actors who had previously given significant thought to the nature of the universe could maintain physiological allostasis during and after an affective crisis by modifying some of their previous epistemological set points. Motivated by the need to accommodate such a modified *Weltmodell*, the subjects of this book were variously driven to alter their views on such diverse topics as the probability of extraterrestrial life, the evolution of the universe and the ages of the earth and sun. It will, for

[15] Thomas Dixon, *From Passions to Emotions: The Creation of a Secular Psychological Category* (Cambridge University Press, 2003), 239–240.

[16] Darwin, *life and letters*, 134.

example, be argued in due course that the Cambridge polymath William Whewell (1794–1866), influenced by the distressing circumstances of the terminal illness of his wife Cordelia (1803–1855, née Marshall), completed an already-nascent metaphysical shift to a position in which the salvation of Man was the primary purpose of creation. In order to bolster this view of final causes, he felt compelled to alter his previously held views about the probability of the existence of extraterrestrial life as well as exhibiting more overtly Christian sentiments in his private conversations and writings.

In an ideal world, such ontological shifts would be easily identifiable in sources made conveniently available by the subjects in question who, despite the negative experiences which they were undergoing, clearly sign-posted the changes in which posterity would be interested by means of a detailed outpouring of their thoughts and feelings. Such records are, unfortunately but unsurprisingly, rarely forthcoming. The relevant accounts which we can locate, as we will discuss further below, are usually retrospective and inevitably couched in terms of concepts and motivations which were suggested by familiar archetypes and which could be seen as painting the writer in a good light. Such narratives were also of necessity couched in terms and using concepts which were familiar to their anticipated audiences as well as their authors. As Boddice and Mark Smith point out, lived experiences are inevitably situated in cultural contexts which supply the raw material, the warp and weft, from which all such accounts are woven. "Culture is everywhere, even in these moments, and the historian's job is to understand culture in its intersection with human brain-body systems."[17]

The importance of the influence of cultural context on scientific productions in particular has been studied by Lorraine Daston, who has convincingly argued that the "doing of science" should be understood as taking place within a "moral economy by which she meant "a web of affect-saturated values that stand and function in well-defined relationship to one another":

> By examining in a new light just those ways of knowing once thought to exempt science from the realm of emotions and values, a study of moral economies may illuminate the nature of the rationality that seemed to exclude them.[18]

[17] Boddice and Smith, *Experience*, 33.

[18] Lorraine Daston, 'The Moral Economy of Science', *Osiris*, (01/1995) 10: 2–24 at 4, 24. Daston points out at the beginning of this article that her definition of 'moral economy' should not be confused with other usages such as that proposed earlier by Edward. P. Thompson.

Otniel Dror et al. have, however, pointed out that there is a particular difficulty in excavating emotions from accounts left by scientists during the nineteenth century given that the evolving moral economies of the era strived to exclude such intangible and subjective elements from their praxis:

> How does the historian identify emotions in a culture that deliberately and methodologically excluded emotions from its texts, narratives, practices, and practitioners [...] This latter difficulty partly explains the relative absence of histories of emotions and histories of science that derive their history of emotions from science qua science[19]

It is also important to note that during the nineteenth century many people who opined on scientific topics, including Whewell, would not have confined themselves to a 'scientific' self but were also wholly or partially engaged in other epistemologies such as, most importantly, the religious "truths without which Faith becomes an absurdity".[20] In other words, the moral economy of science had to contend with parallel moral economies both within the subject and in wider society in some of which subjectivity was more to be valued than objectivity.[21] And astronomy, of course, sat at the very intersection of the competing ontologies offered by theology and evolutionary cosmogony.

Victorian Experiences of Illness and Death

Whewell's shift towards a heightened sense of religiosity during his wife's crisis was not at all unusual. A relatively common experience which constituted a key psychotropic factor in the lives of a number of nineteenth-century physical scientists, as well as the population at large, was the prevalence of life-threatening diseases. Victorian experiences of illness included a variety of syndromes which a later reader may struggle to identify due to differences in the medicalisation of ambiguous conditions and changes in diagnoses, treatments and terminology. Both Cordelia Whewell,

[19] Otniel E. Dror, Bettina Hitzer, Anja Laukötter and Pilar León-Sanz, 'Introduction to History of Science and the Emotions', *Osiris*, 31 (2016), 1–18 at 7–8.

[20] William F. Barry, 'Mr. Tyndall and Contemporary Thought.' *The Dublin Review*, Oct 1876, 27, 54: 431–469 at 469.

[21] Lorraine Daston and Peter Galison, *Objectivity* (New York: Zone Books, 2010), 37.

for example, and Margaret Thomson (1827–1870, née Crum), who was married to the Scottish physicist, William Thomson (1824–1907, later Lord Kelvin), suffered from unnamed, debilitating and ultimately fatal illnesses. They may well represent just two instances of the many nineteenth-century people who suffered from tuberculosis, generally referred to as 'consumption', which was a frequent cause of death at this time and assumed, incorrectly, to be more prevalent amongst 'frail' women.[22] Basil Meyer has summarised contemporary expectations "of those individuals most likely to develop the disease":

> those possessed of a long neck, prominent shoulders or shoulder blades, bright eyes, narrow waist, and fair skin were likely to develop the disease. Such individuals would also be of a sanguine temperament, highly intelligent, sensitive, passionate, and artistic or imaginative. Such ideas were subscribed to by doctors and by the wider public.[23]

Tuberculosis, also known as phthisis or the 'white death', had complex effects not all of which were viewed negatively by contemporary society. The ethereal and attractive paleness which was evident in the appearance of some victims was an attribute which had been considered desirable by Romantic poets and was, for example, the probable cause of the "pale, fragile beauty" of the pre-Raphaelite model Elizabeth Eleanor Siddall (1829–1862). The patient, especially if "more affluent and genteel", could survive for many years during which the disease peaked and declined until it either proved fatal or went into remission. "But neither virtue nor wealth guaranteed protection against tuberculosis, so the disease acquired one of its most persistent attributes: an aura of sacrifice and atonement." In a nineteenth-century context, this attitude of resignation could easily be read as religious affect, a noticeable aspect of the story of Cordelia Whewell.[24]

Both Cordelia and Margaret were indeed affluent and genteel and, assuming they had in fact been afflicted by tuberculosis, were in possession of existing tropes which informed their expectation of the future. As

[22] Katherine Byrne, *Tuberculosis and the Victorian Literary Imagination* (Cambridge University Press, 2011), 30–38.

[23] Basil Meyer, 'Till Death Do Us Part: The Consumptive Victorian Heroine in Popular Romantic Fiction', *The Journal of Popular Culture*, 37, 2, 2003: 287–308 at 289.

[24] Thomas Dormandy, *The White Death: A History of Tuberculosis* (London: The Hambledon Press, 1999), 85–100, 51, 61–62.

Aspasia Stephanou has noted, these were to be found in “the ways nineteenth-century medical and literary discourse facilitated the creation of a pathologized female body”:

> the death of a beautiful consumptive idealized and consumed by the male gaze is represented through the tension between religious imagery and sexual undertones [...] Nineteenth-century medical discourse and literature on consumption exalted the materiality of the consuming female body by transforming suffering into something beautiful, pure and spiritual, or even sexual.[25]

Unlike their working-class counterparts, they would be expected to perform a level of frailty concomitant with the feelings of resignation and an aspiration to the good and apparently, but in practice infrequently, painless death which featured heavily in literary descriptions of this particular disease. Their husbands similarly understood the role expected of them: to support and nurture their wives through to either death or recovery and, in the former case, to encourage both themselves and their partners with the hope of an eventual joyful reunion in the hereafter.[26]

Whether or not an instance of serious illness or injury proved to be fatal, the sufferer and their near friends and relations would quite reasonably assume that death might be lurking just around the corner. This was an era in which even middle- and upper-class British families experienced higher rates of infant mortality and early adult death, primarily due to infectious disease, than is the case for more recent populations in the global north.[27] As a consequence, the expected emotional responses of the soon-to-be departed and the bereaved had been well documented for “the educated English reader”:

> Conventions associated with *ars moriendi* (art of dying), familiar to the educated English reader from Jeremy Taylor’s *Holy Dying* (1651), were deeply rooted in the corporate memory of the faithful, and provided nineteenth-century novelists with excellent material on which to work.[28]

[25] Aspasia Stephanou, ‘Lovely Apparitions and Spiritualized Corpses: Consumption, Medical Discourse, and Edgar Allan Poe’s Female Vampire’, *The Edgar Allan Poe Review*, 14, 1 (Spring 2013), 36–54 at 37–38.

[26] Patricia Jalland, *Death in the Victorian Family* (Oxford University Press, 1996), 39–46.

[27] Thomas Dixon, *Weeping Britannia: Portrait of a Nation in Tears* (Oxford University Press, 2015), 158–159. For a discussion of mortality rates, see: Thomas McKeown and R. G. Record, ‘Reasons for the decline of mortality in England and Wales during the nineteenth century’, *Population Studies* (1962) 16, 2: 94–122.

[28] Michael Wheeler, *Heaven, Hell, & the Victorians* (Cambridge University Press, 1994), 32. Dixon, Weeping, 153–154.

Depending on the religious allegiance of the bereaved, the passing of a loved one, such as the demise of Little Nell in *The Old Curiosity Shop*, could be seen to be an almost-welcome event, bringing relief to their suffering whilst holding out the promise of a passage to a higher plane of existence where they would be reunited with those who had gone before.[29] These expectations were also of great significance to the mortally ill patient. During the first half of the nineteenth century, as Patricia Jalland has pointed out, the influence of the Evangelical wing of the Church of England set the agenda for what a 'good death' should look like, even after the advent of the less febrile influence of Tractarianism. The about-to-be deceased was expected to exhibit a calm resignation together with the signs of salvation, generally in the form of declarations of faith, which were eagerly looked for by their families. Narratives of the deathbed scene were often, consciously or unconsciously, edited to conform to these expectations.[30] It should also be noted, however, that the influence of Evangelical doctrines waned somewhat as the century progressed and the primary concerns of carers began to focus more on the avoidance of physical pain and discomfort.[31]

An important resource which was available to the seriously ill subject was the common stock of conversion stories. These had been popularised by late eighteenth-century religious groups including the Methodists, Evangelical Anglicans and Scottish Calvinists. They generally contained a narrative arc in which a trigger event or the reception of a narrative (such as a sermon or an 'improving' book) was followed by a lengthy period of reflection leading to a declaration of new or renewed faith. Unconversion narratives, on the other hand, were relatively unknown at the beginning of the century but began to be printed later in the period. They detailed a similar lengthy soul-searching process in which, however, a tortured individual eventually comes down on the side of doubt.[32]

[29] Wheeler, *Heaven*, 31–50.

[30] Jalland, *Death*, 21–38.

[31] Jalland, *Death*, 51–58.

[32] Norman Vance, *Bible and Novel: Narrative Authority and the Death of God* (Oxford University Press, 2013), 135–160. Bernard Lightman, 'Robert Elsmere and the Agnostic Crises of Faith.', in Richard J. Helmstadter and Bernard Lightman (eds.), *Victorian Faith in Crisis* (London: Macmillan, 1990), 283–311.

Familial responses to bereavement have also been identified by Jalland and Dixon as having played into those discourses of religious conversion which provided templates for the experience of grief and resolution. Both historians have pointed to, amongst other exemplars, the experiences of Archbishop Archibald Campbell Tait (1811–1882) and his wife Catherine (1819–1878, née Spooner), who responded to the deaths of their children with abundant evidence of grieving but at the same time with a level of stoic acceptance, seeing in these events the mysterious working out of God's plan and assuming that they would in due course be reunited with the departed.[33] Daniel Lord Smail, when making his case for the importance of the intersection of neuroscience and history, argued that such a paradigm shift could also lead to increased performance of the 'outer forms' of the subject's chosen church such that they would, in his view, experience the psychotropic effects of "religion and ritual as sources of dopamine and other chemical messengers"[34]

It is important to bear in mind that conversion in this sense and at this time was not normally a drastic shift from unbelief to belief but rather a change of course within the broader family of Christian churches such as, for example, from a Broad Church or latitudinarian stance to the adoption of an Evangelical confessional alignment. The available templates for a move to a more militant strand of revealed religion set an expectation that the individual would indeed experience a climactic personal conversion, quite possibly in a deathbed setting, and henceforth engage in a more intimate relationship with the deity.[35] Evidence of such a dramatic and climactic event was seen as a necessary precursor to acceptance into the ministry of some nonconformist and Calvinist churches.

The case studies of nineteenth-century people to be detailed in due course will support the argument that the sequence of events which led to the appearance of a conversion or unconversion trope was often prompted and driven by physiological stimuli. One such factor which was particularly prevalent in this period was the relatively widespread therapeutic use of psychogenic substances, primarily opiates, which constitute at least a

[33] Jalland, *Death*, 127–139, Dixon, *Weeping*, 159–162 and Wheeler, *Heaven*, 48–50.

[34] Daniel L. Smail, *On Deep History and the Brain* (University of California Press, 2008), 184–185.

[35] Timothy Larsen, *Crisis of doubt: Honest Faith in Nineteenth-Century England* (Oxford University Press, 2006), 11.

strong suspect in the panoply of affect which impinged on a severely ill or injured patient and which, based on the reported experiences of contemporaneous users, can be asserted to have played a role in the theologically informed responses of nineteenth-century patients.

Psychotropic Drugs in the Nineteenth Century

The use of opium has tentatively been traced back to the Sumerians who are believed to have referred to the poppy as *Hul Gil*—the 'joy plant'. Opiates have had a wide variety of medical uses including as a component of an antidote to poisons prepared for the Emperor Nero (37–68), a soothing medicine for children in mediaeval Persia and as an anti-dysentery drug during the T'ang dynasty in China (618–907).[36] From the early modern period in Europe, they have been used as palliatives, cures for plague, analgesics and sedatives.[37] Opium consumption was widespread throughout the nineteenth century. Even after the Pharmacy Act of 1868, it continued to be widely prescribed as a painkiller, being regarded as less hazardous than some of the other substances controlled by the legislation.[38] It was also during this period that a number of novel surgical anaesthetics were introduced, including ether and chloroform, both of which, like opium, could bring about an altered state of consciousness.[39]

Opium was usually consumed in liquid form, having been dissolved in alcohol to produce a tincture known as laudanum. There were also a number of proprietary preparations which were similarly widely available and

[36] Hooman Khademi, Farin Kamangar, Paul Brennan and Reza Malekzadeh, 'Opioid Therapy and its Side Effects: A Review', *Arch Iran Med.* 2016; 19(12): 870–876, at 870.

[37] Javier Moscoso, 'Exquisite and Lingering Pains: Facing Cancer in Early Modern Europe' in Rob Boddice (ed.), *Pain and Emotion in Modern History* (London: Palgrave Macmillan, 2014), 16–35 at 29–30.

[38] Victoria Berridge, *Opium and the People: Opiate Use and Drug Control Policy in Nineteenth and Early Twentieth Century England* (London: Free Association Books, 1999), xviii–xxi, 64–66, 113–122. The main effect of the 1868 Act seems to have been to develop a form of 'closed shop' for professional pharmacists.

[39] Christopher Partridge, *High Culture: Drugs, Mysticism, and the Pursuit of Transcendence in the Modern World* (Oxford University Press, 2018), 59–87. See also: Guy de Maupassant, 'Dreams' originally published as 'Rêves' in *Le Père Milon* (Paris: Paul Ollendorff, 1899) and Herbert G. Wells, 'Under the Knife', in *Tales of the Unexpected* (London: W. Collins Sons, 1924), 54–70.

affordable.[40] The effects of opium at these dosage levels are in general dependent on two key factors often referred to as 'set' (the individual's previous experience of having taken the drug) and 'setting' (the context in which the substance is being consumed). Furthermore, even if set and setting could be somehow standardised, responses still varied from person to person, eventually resulting, if usage is continued, in addiction for some but not for all. The effects of laudanum consumption were quite mild for many, but for others this psychodysleptic and hallucinogenic concoction would stimulate very vivid dreams during which the subject lost their sense of individual existence. Whether or not such opium dreams are experienced as being pleasant or unpleasant seems to be strongly dependent on the point in the life cycle of usage. Clearly, the pleasurable somnolence-inducing effect of initial exposure is a factor which contributes to subsequent proclivity to long-term addiction.[41]

Opiates were regularly consumed by significant sections of the working-class population. They were also used by a number of writers in the early part of the nineteenth century, often seen as a period of transition from an age of Sensibility, when "the knowledge of oneself, personal identity, stemmed from the sensations represented by imagination", to the advent of Romanticism.[42] The authors in question included Percy Bysshe Shelley (1792–1822), John Keats (1795–1821), Lord Byron (1788–1824), Sir Walter Scott (1771–1832), Branwell Brontë (1817–1848) and Thomas De Quincey (1785–1856). In the case of Shelley, his usage might conceivably have been classed, and was claimed by him, as therapeutic, whereas the consumption of quite prodigious quantities of laudanum by Brontë and De Quincey can only be seen as having been recreational and potentially inspirational in nature at a time when it and similar opiates were believed by many Romantics to be 'stimulants'.[43] "Opium distorted perceptions of time and space and heightened emotional experience, something that strengthened its appeal to the Romantics."[44]

[40] Berridge, *Opium*, 56–57. Louise Foxcroft, *The Making of Addiction: The "Use and Abuse" of Opium in Nineteenth-Century* Britain (Aldershot: Ashgate, 2006), 11.

[41] Partridge, *High Culture*, 8–29.

[42] Daston and Galison, *Objectivity*, 208. Rachel Hewitt, *A Revolution of Feeling: The Decade that Shaped the Modern Mind* (London: Granta, 2017), 68–69, 293–295.

[43] Berridge, *Opium*, 21–37, 49–61.

[44] Richard J. Evans, *The Pursuit of Power: Europe 1815–1914* (London: Penguin, 2016), 449.

Opium was explicitly resorted to as an inspiration by some later authors. Both Wilkie Collins (1824–1889) and Edgar Allan Poe (1809–1849) noted that there is a particular phase in the effects of opium on the mind which is conducive to the creation of imaginative plotting, sometimes accompanied by the onset of the horrific creations associated with the liminal region between life and death. The results of such morbid ruminations can be seen in both Poe's Gothic fantasies and more explicitly in Collins' *The Moonstone* in which laudanum not only formed the author's inspiration but also acted as a key plot device.[45] Such fatalistic and doom-laden visions would have been of particular significance to subjects who were indeed physically as well as mentally on the threshold, as they then saw it, of their own mortality.

Consumption of psychotropic substances, given the right set and setting, can also act as a catalyst for feelings of religiosity and sublimity.[46] This association between narcotic euphoria and potential religious affect was noted by one of the earliest psychologists, William James (1842–1910), author of a weighty tome entitled *Varieties of Religious Experience* (1902), in which he prioritised the transcendental thoughts and emotions of the subject over the 'outer trappings' and formal structures of religion. James identified a significant connection between such feelings of transcendentalism, which he interpreted as religious experiences, and the use of "mind-altering drugs", including his own experience of exposure to nitrous oxide.[47]

[45] Sadie Plant, *Writing on Drugs* (New York: Picador, 2001), 24–32.

[46] For a more extensive discussion of the connections between psychoactive drugs and religious experiences see, for example: Aldous Huxley, *The doors of perception; and, Heaven and hell* (London: Chatto and Windus, 1956) and Plant, *Writing*, 93–118. Double-blind trials with psilocybin, a potent hallucinogen which has been used for sacramental purposes in a number of cultures, have shown that when administered to volunteers who reported regular participation in religious or spiritual activities, it "occasioned experiences similar to spontaneously-occurring mystical experiences that, at 14-month follow-up, were considered by volunteers to be among the most personally meaningful and spiritually significant of their lives". See: Roland R. Griffiths, William A. Richards, Matthew W. Johnson, Una McCann and Robert Jesse, 'Mystical-type experiences occasioned by psilocybin mediate the attribution of personal meaning and spiritual significance 14 months later', *Journal of Psychopharmacology*, 22, 6 (2008): 621–632 at 621. https://doi.org/https://doi.org/10.1177/0269881108094300.

[47] William James, *The Varieties of Religious Experience: A Study in Human Nature; Being the Gifford Lectures on Natural Religion Delivered at Edinburgh in 1901–1902* (New York: Longmans, Green, 1902), 387–389.

In *Confessions of an English opium-eater*, De Quincey's initial experience of 'opium eating' was narrated in the form of a conversion story in which, as noted by Christopher Partridge, he likened the effect of the narcotic to attendance at a new church—the "church of opium":

> The opium-eater, he declared, "feels that the diviner part of his nature is paramount; that is, the moral affections are in a state of cloudless serenity; and overall there is the great light of the majestic intellect." This, says the enthusiastic convert to psychedelic gnosis, "is the doctrine of the true church on the subject of opium: of which church I acknowledge myself to be the only member—the alpha and the omega."[48]

De Quincey also noted that the effects for him of opium consumption included significantly distorted perceptions of time and space which he described as the opening up of a "theatre […] which presented nightly spectacles of more than earthly splendour":

> Space swelled, and was amplified to an extent of unutterable infinity. This, however, did not disturb me so much as the vast expansion of time; I sometimes seemed to have lived for 70 or 100 years in one night; nay, sometimes had feelings representative of a millennium passed in that time, or, however, of a duration far beyond the limits of any human experience.[49]

It is evident therefore that a significant number of nineteenth-century artists and writers, in common with many other members of the population, regularly resorted to opium in its various forms, for a mixture of medical, recreational and inspirational reasons. "Many, besides De Quincey, have written in eloquent praise of 'the abyss of divine enjoyment' which these drugs reveal."[50] Like the "Lotos-Eaters" described by Alfred Lord Tennyson (1809–1892), in a poem probably written as a cautionary tract on the negative effects of opium eating which he had observed amongst close family members, they from time to time languished in states of lethargy whilst, at least initially, experiencing sybaritic visions infused

[48] Partridge, *High Culture*, 41. Thomas De Quincey, *Confessions of an English opium-eater*, Second Edition (London: Taylor and Hessey, 1823), 98.

[49] Ibid., 158–160.

[50] Mahabir Winston, 'History of the Psychedelic Drug Experience', *Canadian Psychiatric Association Journal*, 13, 2 (1968): 189–190.

with distorted perceptions of space and time accompanied by the dissolution of the self.[51]

The consequences of opiate use, whether deliberately sought out or occurring as a side effect of its employment as an anaesthetic, analgesic or anodyne (in the pre-twentieth-century sense of a substance which reduces pain), can consequently be suspected as having played a part in several of the significant epistemological changes to be discussed in due course. We will note, for example, in the case of William Thomson that he was subjected to a mixture of psychotropic substances, including opiates and chloroform, during a serious injury and at a time when we can reasonably assume that he came to understand his own body to be a site which demonstrated the inevitability of increasing entropy. This personal example of the cosmic dissipation implicit in his new thermodynamics was followed by a deeper engagement with questions about the scale of time and the nature of the universe. Such a contingency cannot be absolutely proven retrospectively, but the circumstantial evidence for such a linkage will be shown to be worthy of further investigation.[52]

Victorian Religion

A topic which has attracted much historiography and which is relevant to the shifting sands of scientific thinking, with particular reference to the subjects of this book, is the evolving nature of Victorian religion and its relationship with scientific thinking. The inference will be drawn that the contemporary context of relevant theologies informed the exact nature of the conversion or unconversion experienced by the actors in question, subsequently shaping their specific beliefs about the natures of both the deity and the universe and lent shape, colour and conviction to their public and private utterings.

The first half of the nineteenth century was characterised by a series of tensions between contending strands within the established churches in

[51] Alfred Lord Tennyson, 'The Lotos-Eaters' in Karen Hodder (ed.), *The Works of Alfred Lord Tennyson* (Ware: Wordsworth Editions, 1994), 93–97. There are reasonable grounds to suppose that the members of Odysseus' crew who sojourned in the land where "the poppy hangs in sleep" (l. 56) reflect Tennyson's concerns about the negative effects of opium eating which he had observed in his father and brother. See: Catherine B. Stevenson, 'The Shade of Homer Exorcises the Ghost of De Quincey: Tennyson's "The Lotos-Eaters"', *Browning Institute Studies*, vol 10 (1982): 117–141.

[52] Boddice, *History*, 39–40.

Britain and Ireland. Churches in England were influenced by the ongoing fallout of the eighteenth-century Evangelical movements whose members were more radical and proselytising than their more passive and conservative peers. In assessing the influence of Evangelicalism on Victorian attitudes to death, Jalland has observed:

> Evangelicalism was a vital force in the development of Methodism, and it transformed the Church of England, as well as influencing such dissenting sects as the Congregationalists, the Baptists, the Presbyterians, and even the Quakers. Evangelicalism touched all denominations and all sections of the Church of England, including even the High Church.[53]

In the 1830s and 1840s, there were further potential and actual divergences. Emanating from the University of Oxford, the Tractarians argued for a return to traditional forms of liturgy which were seen by others as veering too much towards Catholicism. In 1843, the Church of Scotland, a Presbyterian and strongly Calvinist institution which was broadly subdivided into Moderate and Evangelical wings, underwent a major schism when the Scottish theologian and political economist Thomas Chalmers (1780–1847) led a breakaway by a substantial minority of Evangelist clerics. They founded the Free Church of Scotland, known vernacularly as the 'wee frees', in which, in keeping with traditional Presbyterian ideology, the congregations regained overall control of the appointment of their ministers. Divisions also occurred in Ireland, where the established church saw a move away from the Anglo-Irish flavour of Protestantism towards a greater recognition for traditional Catholicism in the south and a more entrenched, originally Scottish, variety of Calvinism in the north.[54]

From the beginning of the century onwards, both the established churches and their nonconformist peers were also beginning to perceive that traditional Christian beliefs were facing two significant external challenges. Firstly, the conclusions to be drawn from recent astronomical and geological observations were increasingly difficult to reconcile with the Mosaic chronology found in the book of Genesis. Secondly, the emerging praxis of textual analysis in Germany cast significant doubts on the inerrancy and traditional authorship of the Bible. Doubts occasioned by this 'higher criticism' were not yet widespread amongst the lay public, but the

[53] Jalland, *Death*, 19–20.

[54] William Hanna, *Memoirs of the Life and Writings of Thomas Chalmers D.D., LL. D.*, vol IV, (Edinburgh: Thomas Constable, 1852), 344–364. K. Theodore Hoppen, *Ireland since 1800: Conflict and Conformity* (Harlow: Longman, 1999).

putative findings of this philological process were already known to the more educated and widely read members of the clergy.[55]

In response to the chronology challenge, a number of commentators sought to downplay the findings of science with regard to the scale and duration of creation in the early decades of the nineteenth century. One species of response took the form of "passionately immoderate Biblicists committed to the reconstruction of Mosaic cosmogonies and geologies preserving six-day Creation and Flood".[56] A more reasoned and authoritative response to recent controversies was a series of publications sponsored by the Earl of Bridgewater (1756–1829) which promoted the concepts of natural theology. The *Bridgewater Treatises* followed a line of argument perhaps most notably expounded by William Paley (1743–1805) in *Natural Theology or Evidences of the Existence and Attributes of the Deity* (1802).[57] In this widely read work, Paley, building on the work of a number of Enlightenment thinkers, developed the concept of God as a blind watchmaker who created the universe and set it in motion in accordance with a set of immutable and divinely ordained laws. In Paley's view, this deistic proposal, far from negating the concept of an omnipotent creator, merely pointed up the very considerable powers of design and construction possessed by the prime mover.

When suggestions originated from Germany that the Bible may not be the unalloyed word of God but may rather have been the result of the efforts of a number of human authors, they resulted in some febrile controversies. The most notable followed the 1860 publication of an edition of *Essays and Reviews* in which a number of Anglican scholars reported with apparent approval the results of the higher criticism. This religious provocation followed hard on the heels of the publication of Darwin's *On the Origin of Species* in 1859. Later in the century, this coincidence of challenges was one of the factors which allowed two American academics (John William Draper (1811–1882) and Andrew Dickson-White (1832–1918)) to suggest that these two epistemologies had been and were still actually at war, although their motivation for saying so may have

[55] Frank M. Turner, The Victorian Conflict between Science and Religion: A Professional Dimension, *Isis*, 69, 3 (Sep., 1978): 356–376 at 357. Gerald Parsons, 'Biblical Criticism in Victorian Britain: From Controversy to Acceptance?' in Gerald Parsons (ed.), *Religion in Victorian Britain: Volume II Controversies* (Manchester University Press, 1988), 238–257.

[56] Parsons, 'Criticism', 244.

[57] William Paley, *Natural Theology or Evidences of the Existence and Attributes of the Deity Collected from the Appearances of Nature* (Philadelphia PA: John Morgan 1802).

been not entirely unconnected with a need to establish the professional status of science teaching at that time.[58]

There is, on the other hand, some significant evidence that the so-called 'crisis of faith' in the second half of the nineteenth century has been overstated. As Gowan Dawson has pointed out at length, Darwin's thesis was neither as controversial nor as novel, as has sometimes been suggested.[59] Ideas of species transformation had been 'in the ether' since the days of Jean-Baptiste Lamarck (1744–1829), albeit he was a deist tainted in British eyes by the events of the French revolution. Similar concepts had been proposed by Darwin's own religiously heterodox grandfather Erasmus Darwin (1731–1802) and had been expounded at length in a popular but controversial treatise entitled *Vestiges of the Natural History of Creation* (1844).[60]

It is also not clear to what extent that there was a substantial decline in religiosity in general. Timothy Larsen, for example, has argued that the increasing number of 'loss of faith' narratives in the latter part of the period may be more of a reflection of what was an exceptional occurrence and that there was also a significant number, less noticed in recent literature, of reconversions in the same period.[61] Despite these reservations, however, there was, from mid-century onwards, a slow movement away from the established churches and towards new belief systems which sought to provide alternative evidences of eternity with which to replace the partially discredited traditional theologies. These included the advent of spiritualism and the formation of a number of new societies which combined apparently ancient wisdom with religious and metaphysical themes such as the Theosophists and the Hermetic Order of the Golden Dawn.[62]

[58] John W. Draper, *History of the Conflict between Religion and Science* (New York: Appleton, 1875). Andrew D. White, *A History of the Warfare of Science with Theology in Christendom*, 2 volumes (London: Macmillan, 1897). For a discussion of the political purposes of the hypothesised conflict, see: Peter Harrison, "'Science' and 'religion': Constructing the boundaries", in Thomas Dixon, Geoffrey Cantor & Stephen Pumfrey (eds.), *Science and Religion: New Historical Perspectives* (Cambridge University Press, 2010), 23–49 at 27.

[59] Gowan Dawson, *Darwin, Literature & Victorian Respectability* (Cambridge University Press, 2007).

[60] See, for example, Paul Elliott, 'Erasmus Darwin, Herbert Spencer, and the Origins of the Evolutionary Worldview in British Provincial Scientific Culture, 1770–1850', *Isis*, 94, 1 (March 2003): 1–29. Robert Chambers, *Vestiges of the Natural History of Creation* (London: John Churchill, 1844).

[61] Timothy Larsen, *Crisis of Doubt: Honest Faith in Nineteenth-Century England* (Oxford, 2006).

[62] Alex Owen, *Place of Enchantment: British Occultism and the Culture of the Modern* (University of Chicago Press, 2004).

In some cases, these can be seen as groups which fostered a form of Gnosticism in which the initiates are made privy to the inner secrets and the promises of mystical powers which are not to be revealed to the unenlightened masses in the outer world. In the case of spiritualism, however, there was clearly an effort to counter the threatening findings of secular science with a new and empirically founded basis for belief in immortality based on evidence garnered from the ethereal inhabitants of the afterlife. In due course, it was garnished with a coating of respectability and rationality by the formation of the Society for Psychical Research in 1882 with a strong leavening of well-known physicists, philosophers and politicians amongst its early leaders.[63]

The relationship, therefore, between Victorian science and religion clearly reflected a complex set of issues which not only afflicted society at large but were also played out within the *Weltmodelle* of the subjects whose work and views are to be covered in succeeding chapters. A slow shift during the nineteenth century from public to private religion can be glimpsed in the changing responses to affect-laden crises. The Evangelical conversion exhibited by Chalmers, a corporate Calvinist, can be contrasted with the initially semi-detached Anglicanism of Whewell which *in extremis* elided into a profession of increased faith. A little later still, the popular astronomer Richard Anthony Proctor (1837–1888) defined his prevaricating personal position as much by what he was opposed to as to what he advocated, managing to switch with relatively little difficulty from Anglicanism to Catholicism and thence to agnosticism. A similar timescale applied to claims of authority in the interpretation of natural law which ranged from the creationist certainty of clerics like Paley to later assertions of scientific rationality made by agnostics, pantheists and other holders of heterodox views. In the 1860s and 1870s, there were nonetheless some theologically nuanced responses to nascent materialism which emanated from professional scientists whose 'religious' selves had absorbed elements of early exposure to protestant enthusiasm.[64]

[63] Jalland, *Death*, 365–366. Janet Oppenheim, *The Other World: Spiritualism and Psychical Research in England, 1850–1914* (Cambridge University Press, 1985).

[64] Frederik Gregory, 'Intersections of Physical Science and Western Religion in the Nineteenth and Twentieth Centuries' in Mary J. Nye (ed.), *Cambridge History of Science*, vol 5, chapter 2 (Cambridge University Press, 2003): 36–53 at 39–44. Crosbie Smith and M. Norton Wise, *Energy and Empire: A Biographical Study of Lord Kelvin* (Cambridge University Press, 1989). Bernard Lightman, 'Victorian Sciences and Religions: Discordant Harmonies', *Osiris*, 2nd Series, 16, Science in Theistic Contexts: Cognitive Dimensions (2001): 343–366.

Subjects and Sources

Clearly, the argument being made here could and should be broadly applicable, whilst being dependent on cultural contexts for its specific details, to any group of people at any time in history but this monograph will focus on the nineteenth century in Britain and Ireland. This was a period of particular dialectical contention between the overlapping epistemologies of religion and science.[65] Given that the historiography of geology and biology and the ways in which they were interpreted at that time in order to support or oppose developmental processes which could be deemed to be consistent with theological worldviews is relatively voluminous, we will be concentrating instead on the views of a subset of those natural philosophers who opined on the structure, size and composition of the universe itself. Whilst most of the ensuing discussions will be concerned with objects and concepts which operate at cosmic scales, the discussion of the emerging role of energetics in the second half of the century will necessarily introduce some parallel discussions of the nature of matter at very small scales.

Astronomy had made and was making significant advances during this period. By the end of the eighteenth century, the telescopic investigations of Frederick William Herschel (1738–1822), Charles Messier (1730–1817) and others had determined that our sun is a very average body which, rather than being static as previously supposed, was in constant motion across a vast amorphous sea of stars.[66] Thus, astronomers were wrestling with changing conceptions of an expanding cosmos which, whilst inspiring a sense of awe on the one hand, also bore potential harbingers of religious doubt on the other.

We shall examine relevant excerpts from the lives of some significant contributors to nineteenth-century astronomy and physics to identify the key cultural contexts and adventitious affects which influenced them and to see how these inputs played out in the selection and presentation of their ideas. One of the key characteristics in the lives of our 'few' chosen individuals who were privileged to have voices which have been preserved

[65] See, for example, Frank M. Turner, 'The late Victorian conflict of science and religion as an event in in nineteenth-century intellectual and cultural history', in Thomas Dixon, Geoffrey Cantor and Stephen Pumfrey (eds.), *Science and Religion*, 87–110.

[66] Michael A. Hoskin, 'William Herschel and the Construction of the Heavens', *Proceedings of the American Philosophical Society*, 133, 3 (Sep., 1989): 427–433.

in written form is that unlike the 'many' who were their contemporaries and predecessors, they also had the opportunity to inhabit more than one cultural context, being in most cases engaged in overlapping disciplines which were at that time usually taken to be complementary rather than competing. One important overlap which applied to most of our subjects was the intersection between religion and higher education. In the early nineteenth century, many graduates with an interest in natural philosophy had also been ordained in one of the established churches. As Frank Turner put it:

> Since the seventeenth century the parson-naturalist and the academic clergyman-scientist had played a major and by no means inglorious role in British science, as the names of John Ray, Joseph Priestley, John Stevens Henslow, Adam Sedgwick, and William Whewell attest.[67]

An important resource derived in part from their experiences in their respective churches and which was of great value to several subjects in staking a claim to authority and authenticity was that of public performance. We learn from contemporary sources that some of our actors were equipped with performative selves which were able to generate or simulate the external evidence of their emotions and thus provoke an intended response in their audiences.[68] John Pringle Nichol (1804–1859), for example, was a Scottish professor of astronomy who was capable of attracting and holding the attention of large publics, both in person and in print. His success was clearly the result of an ability, established at an early age and aided by his earlier experience of preaching, to fascinate and inspire others by virtue of the affective verbal and non-verbal grace notes which accompanied his expansive explanations of the 'new astronomy'.[69]

[67] Turner, 'Victorian Conflict', 364. For a Scottish perspective, see also: Crosbie Smith, *The Science of Energy: A Cultural History of Energy Physics in Victorian Britain* (London: Athlone, 1998), 5.

[68] Clayton D. Drinko, *Theatrical Improvisation, Consciousness, and Cognition* (New York: Palgrave Macmillan, 2013). Boris C. Bernhardt, and Tania Singer, 'The neural basis of empathy', *Annual review of* neuroscience, 35 (2012): 1–23.

[69] See discussion of Nichol's success as a popular astronomer in Chap. 8. Nichol represents somewhat of an exception to the majority of our subjects in that, probably because of the limited amount of available correspondence, no complete account of his life has been published to date.

The scientists whose stories will be explored in this book were recruited from that section of the population who could either afford a university education or whose early progress was such that they gained sponsorship which moved them into that numerically limited and fortunate position. Women on the whole were not able to access similar levels of education or subsequent professional vocation. There were some notable exceptions, however, including the three women who were elected honorary members of the Royal Astronomical Society during the nineteenth century. In 1903, Margaret Lindsay Huggins (1848–1915) and Agnes Mary Clerke (1842–1907) similarly attained honorary membership, but full membership for women was not available until 1915.[70]

There is thus a limited pool of female astronomers who could have been selected for inclusion in this monograph. One candidate, Clerke, appears to have had an interesting career combining an Irish Catholic background with an early interest in astronomy. She contracted a serious illness which began in childhood and led to her spending most of her time between the ages of 25 and 35 in Italy. It would be interesting to know to what extent the divine ontology and theological awe which informed her widely read interpretations of the discoveries of astronomers and descriptions of the physical universe were influenced by her early maladies. Unfortunately, like most women of this era, contemporary sources for her life are very limited: there being only a brief *Appreciation* written by Huggins and privately circulated in 1907, shortly after Clerke's death.[71]

The main route by which we can endeavour to understand the interiority of our subjects is to examine those public and private written sources which are still available. In some cases, authors are believed, quite understandably, to have carefully crafted their apparently private commentaries in order to cohere with a pre-existing public persona. One example of such a writing strategy is the passive voice and inert style adopted by

[70] The three women who attained honorary membership in the nineteenth century were: Caroline Herschel (1750–1848), Mary Somerville (1780–1872) and Caroline Sheepshanks (1789–1876). John L. E. Dreyer and Herbert H. Turner, (eds.), *History of the Royal Astronomical Society: Volume 1 1820–1920* (1923 repr. Oxford: Blackwell, 1987), 81, 192, 234. Bernard Lightman, 'Constructing Victorian heavens: Agnes Clerke and the "New Astronomy"' in Barbara T. Gates and Ann B. Shteir (eds.), *Natural Eloquence: Women Reinscribe Science* (University of Wisconsin Press, 1997), 61–75.

[71] Margaret L. Huggins, *Agnes Mary Clerke and Ellen Mary Clerke: An Appreciation* (Printed for Private Circulation, 1907).

Darwin as a strategy to conceal some of his potentially more controversial ideas about religion from potential biographers.[72] To quote John Hedley Brooke:

> Darwin disliked interrogation, protesting that he could not see why his beliefs should be of interest to anyone but himself. The complication here is that his writings did contain remarks calculated to cause least offence. He knew there were things he should say and not say, particularly concerning the human mind, if he wished to retain public sympathy. […]. The upshot is that there are degrees of ambiguity in Darwin's remarks about religion that can make them difficult to interpret.[73]

It is generally true, in any case, that a personal memoir or diary records a set of moments of time which have been selected, reordered, expanded or contracted to suit the author's purposes. All of the available sources have undergone such a process of selection, both in terms of recall and in the editing of the material. It is also inevitably the case, as other scholars have pointed out, that historical actors can no more describe their inner feelings *in vacuo*, that is, in the absence of cultural context and preceding experience, than can we.[74]

Another quite understandable obstacle to a closer understanding of our subjects is that most people, including those who are normally quite communicative, of necessity tend not to produce many letters or diary entries at times of severe illness or during the early phases of convalescence. Indeed, this is a general problem for historians: at times of great stress or during what we have subsequently come to regard as a crucial point in time, the creation of contemporary written records which exactly address our *post hoc* questions is a comparatively rare occurrence.[75] We are consequently compelled, in our attempts to reconstruct their lived experiences, to examine the recollections of family and friends in addition to the kind

[72] John Sturrock, *The Language of Autobiography: Studies in the First Person Singular* (Cambridge University Press, 1993), 215–216.

[73] Brooke, 'Darwin', 394.

[74] Boddice and Smith, *Experience*, 30–33.

[75] See, for example, Ronald Hutton, *The Making of Oliver Cromwell* (New Haven CT, Yale University Press, 2021), 32–47. Hutton, having examined the available sources and hypotheses, concludes that while it seems certain that Cromwell underwent a self-described conversion experience which modified his *Weltmodell* and led to changes in his religious and political inclinations, it is impossible to now say exactly when this change happened or what caused it.

of edited memoir just discussed. And these parallel accounts, which we hope will provide an orthogonal perspective are, of course, just as much environmentally conditioned as are the productions of the principal actors, often being based on somewhat formulaic tropes which we tend to regard, from within the cocoon of our own moral economies, as being pious genuflections without being able to truly judge the extent to which the writers genuinely experienced the apparently conventional sentiments they were expressing. In other words, we need to be careful to avoid the solipsistic practice of imposing our own emotions, conditioned by our *Weltmodelle*, onto the lives of nineteenth-century people. Our best evidence for invisible interior affects may indeed be the externally observable effects thereof.

Structure of This Book

The main body of this book is divided into three parts. Part II will focus on pluralism, which in this context means the question of the possible existence of life, especially where that life consists of intelligent beings, on other planets in the solar system and elsewhere. For Trinitarian theologians, such potential additions to God's flock raised many interesting questions, such as whether extraterrestrial beings have souls and whether they had been redeemed by a single Incarnation on earth, or were multiple redemptive acts required? Did the existence of other beings increase or decrease the evidence for the existence of a deity, and what did it mean for the veracity of the Christian religion? Since much of this debate revolved around imaginative interpretations of limited observational evidence, the participants were relatively free to elaborate their hypotheses and to project their ideas in such a way as to support their personal preferences.

Evidence of extraterrestrial life was variously seen as an argument either for or against such concepts as theistic religion, hierarchies of creation, societal evolution and even as a pattern for imperial conquest and colonisation. Pluralist concepts could be interpreted as metaphors for ideas of racial difference which had inspired a parallel debate between proponents of Christian universalism, albeit harbouring strong paternalistic overtones, and the supporters of human evolutionary hierarchies "with white Anglo-Saxons at the apex".[76]

[76] Steven J. Dick, 'Cosmic Evolution: History, Culture, and Human Destiny' in Steven J. Dick and Mark L. Lupisellsa (eds.), *Cosmos & Culture; Cultural Evolution in a Cosmic Context* (Washington DC: NASA, 2009), 29–59. Catherine Hall, *Civilising Subjects: Metropole and Colony in the English Imagination 1830–1867* (Cambridge: Polity, 2002), 17.

Three of the participants in the pluralism debate were Thomas Chalmers, William Whewell and Richard Proctor. In each case, it will be argued that there is at least circumstantial evidence of a relationship between changes in their interpretation of the likelihood of the existence of extraterrestrial life and preceding or parallel episodes of emotional upheaval and the heightened prominence or reformation of their religious selves.

Part III will examine the stories of several subjects who were engaged in a debate about the origins of the universe in the form of the two cosmogonic hypotheses which had been articulated by Pierre-Simon Laplace (1749–1827) and William Herschel at the end of the eighteenth century. This discussion was given fresh impetus and a wider audience by some of John Pringle Nichol's writings in the 1830s and by the publication of *Vestiges* in 1844. When viewed alongside preceding and contemporary debates in the field of geology, these nebular hypotheses provided additional support for longer chronologies and a developmental history of the cosmos.

They key events in this history include the claimed refutation of the nebular hypothesis based on observations made with a giant telescope located in the midlands of Ireland: the 'Leviathan of Parsonstown'. There was, however, a degree of divergence between two of the key players in the management and reporting of the results of these observations. These differences can be seen to reflect a dialectical process associated with the emergence of 'mechanical objectivity', a moral economy identified by Daston and Peter Galison as an 'epistemic virtue' particularly associated with the state of development of the scientific self in the middle of the nineteenth century.[77] Nichol's response to the two Irish observers endeavoured to subtly undermine their claimed negation of nebulosity and to lay a strong foundation for the use of astronomy as a tool to support theories of cosmic development which were extensions of his earlier support for radical ideas of the political evolution of human society.[78]

The fourth part will examine the debate about the age of the solar system which emerged in the second half of the nineteenth century. This

[77] Daston and Galison, *Objectivity*, 17–19, 39–42, 191–252.

[78] Simon Schaffer, 'The nebular hypothesis and the science of progress' in Moore (ed.), *History*, 131–164.

interchange of views added fuel to existing arguments about the reliability or otherwise of geological and biological chronologies. The subtext of such debates was a form of existential angst: "one of the key factors connecting thermodynamics and astronomy is that they raise similar questions and anxieties about the role of humanity in the universe, and humankind's eventual place in the afterlife."[79]

The main proponents of a low chronology were the 'North British' physicists. Crosbie Smith and Norton Wise, in their comprehensive treatise on the development of nineteenth-century ideas about energy, have eloquently demonstrated the associations between the theological contexts of this group of scientists and their thermodynamical theories.[80] This section will build on their work by identifying several significant life events which affected the same group of physicists and, it will be argued, may have been instrumental in producing contingent shifts in their subsequent choices of research topics and publications. They persisted in holding a belief in a potentially theistic cosmos which was clearly a factor in their advocacy of a young sun and earth, as well as offering a number of speculative ideas about such metaphysical topics as the locus of free will, the existence of the soul as a separate entity and the possibility of eternal life. An additional point to be developed in this section is that to treat them as a homogenous group of Calvinist scientists is to oversimplify the situation: they were perhaps more united by geography and institutional affiliation than they were by confessional alignment or extent of religiosity.

Astronomic Emotions

In the course of studying the narratives of some of the nineteenth-century people who contemplated the nature of the universe, it will become clear that they and their biographers occasionally referred to significant life crises which preceded important changes in their theological, metaphysical and scientific worldviews. It will be argued throughout this book that the impact of such a crisis is to drive a shift in the subject's *Weltmodell*. The timing and extent of the resulting change depended on the severity of the stimulus and the closeness of the object of an event to the individual in question: threats to the continuing existence of the subjects themselves

[79] Niamh Brown, 'Devotional Cosmology: Poetry, Thermodynamics and Popular Astronomy, 1839–1889', PhD thesis submitted to the University of Glasgow (2016): 47.

[80] Smith and Wise, *Energy*. Smith, *Science*.

being self-evidently the most influential. Subsequent adjustments, in some but not all cases, involved a modification of the narratives which the subject used to explain the nature of the universe and, perhaps more importantly, the probability of survival of the soul. It will also become clear that the sites of trauma in the nineteenth century were sometimes accompanied by exposure to psychotropic substances which can perhaps best be understood as externally applied analogues of the internally generated molecules which drive, at least in part, the responses of the brain–body system.

These arguments need, of course, to be pursued with some care. It is important to consider the available evidence 'in the round' when assessing the relative weightings which should be given to the various factors which influenced a subject's *Weltmodell*. Thus, for example, the cases of Thomas Chalmers and John Pringle Nichol will demonstrate a linkage between life-threatening illnesses and religious conversion or unconversion, respectively. Their subsequent forays into astronomical themes provide evidence of at least a partial chain of causality leading from their modified theological leanings towards their emergent cosmologies.

Some potential connections are even subtler. The theological and astronomical reactions of William Whewell to his wife's terminal illness and intimations of her mortality were not instantaneous events but, like her physical decline, developed over a period of several years. It is presumably important in Whewell's case that he was experiencing a lengthy period of anxiety and sympathy for his wife rather than suffering an immediate personal tragedy.

Other narrative arcs, such as several of those described in Part IV, include significant time lapses between affect and effect. In these cases, it is more difficult to be clear that there was a strong causal link between the events in question, as opposed to the influence of the injury or illness being one of several physical and cultural factors which led to a shifting set of priorities. This less direct interplay will be argued in the cases of Henry Charles Fleeming Jenkin (1833–1885) and Balfour Stewart (1828–1887), who responded to a combination of causes, including illness, injury and assaults on their theologically informed science, with attempts to undermine Darwinian chronology (Jenkin) and imagined eternal universes (Stewart).

These less obviously contingent cases further illustrate the more general point that relatively minor affects can make it difficult for our biocultural brains to perform the epistemic virtue of objectivity when faced with perceived challenges to the integrity of the self. It is demonstrably difficult for us to ignore the weight we place on the outcome of any series of events which have implications for our allostatic load or even to objectively assess less impactful options which nonetheless threaten to perturb our *Weltmodelle*.[81] This is not to suggest, however, that the idea of aspiring to objectivity should be abandoned but rather that we need to ensure that the cognitive biases which frequently come into play are more widely recognised and acknowledged as key components of any knowledge-making process, including the past or present formation of scientific ideas and concepts.

[81] For a general review of the concept of confirmation bias, see: Margit E. Oswald and Stefan Grosjean, 'Confirmation Bias' in Rüdiger F. Pohl (ed.), *Cognitive Illusions: A Handbook on Fallacies and Biases in Thinking, Judgement and Memory* (Hove: Psychology Press, 2004): 79–86.

PART II

The Extraterrestrial Life Debate

CHAPTER 2

Planets and Pluralism: How Many Revelations Are Required to Redeem the Entire Universe?

Monarch of Gods and Daemons, and all Spirits
But One, who throng those bright and rolling worlds
Which Thou and I alone of living things
Behold with sleepless eyes!

Percy Bysshe Shelley, 'Prometheus Unbound', Act 1, Scene 1[1]

The concept of pluralism (in the sense of a belief in the existence of extraterrestrial life) had a long prehistory prior to the nineteenth century. Debates about the possible nature of such hypothetical organisms have continued intermittently until the present day fuelled by discoveries in the field of astronomy. It could be argued that the nineteenth century represented a significant peak in the occurrence of such discourses, beginning with the promulgation of pluralist ideas by the proponents of natural theology and ending with a widely read and vigorous debate about the possible existence of artificial canals on the planet Mars. The purpose of Part II, however, is not to examine the case for or against the existence of extraterrestrials but rather to examine the theological, metaphysical and cultural origins of the specific stances on this question which were adopted by some of the protagonists.

[1] Percy B. Shelley, *Prometheus Unbound: A Lyrical Drama in Four Acts with Other Poems* (London: C and J Ollier, 1820), 19.

H. Carlton, *Cosmology and the Scientific Self in the Nineteenth Century*, https://doi.org/10.1007/978-3-031-05280-4_2

In due course, we will see that the views of several subjects with respect to the idea of pluralism were malleable and *Weltmodell* dependent.[2] Such opinions were shaped, sometimes significantly, by the results of interactions between longer-term cultural inputs and shorter-term psychological and physiological affects. The exposition of this topic will necessarily entail a parallel discussion of why, particularly in the nineteenth century, the question of whether or not aliens existed and their hypothesised states of redemption or 'grace' were topics of great interest to both clerics and astronomers, despite the comparative lack of either relevant scriptural sources or observational evidence.

Unlike some other philosophical controversies aired in this period, such as the debates surrounding the limited age of the Earth suggested by Mosaic chronology, belief or otherwise in the existence of extraterrestrial life did not have a clear-cut connection to personal positions in other epistemologies including religion and metaphysics. This disconnect occurred because the various hypothesised alien 'states of being' were easily capable of being adjusted to fit each subject's endeavour to create a narrative of how the universe is ordered from both a theological and a cosmological perspective.

The Origins of Pluralism

Pre-classical people held a variety of beliefs about associations between the planets or stars and various divine beings such as the gods of the Babylonian and Egyptian pantheons as well as, in the latter case, the translated souls of dead pharaohs. Several Greek philosophers including Leucippus (fifth century BCE), Democritus (c. 460–c. 370 BCE) and Epicurus (341–270 BCE) subsequently promoted the idea that the wandering planets are inhabited by other beings. The thoughts of Epicurus on this and many other philosophical topics were subsequently included in an influential work by the Roman poet and philosopher Titus Lucretius Carus (99? BCE–55? BCE) which argued for the concept of a universe which could develop in and of itself without a requirement for divine intervention. These early polytheistic and pluralist ideas were subsequently replaced in western Europe by Aristotelian and Platonic assumptions of the Earth's

[2] A German word meaning 'world picture' which has previously been employed to describe both the cosmological models of physical scientists and the creation stories found in traditional religions. It thus encompasses the two key epistemologies under discussion here.

uniqueness within a specifically Christian conception of the cosmos in which heaven and hell were generally believed to be real locations existing in the pre-Copernican universe of the late classical and mediaeval periods which consisted of a series of spheres and which were thought to be populated by a diverse collection of angels, demons, saints and sinners.[3]

The astronomical observations made in the late eighteenth century by Frederick William Herschel (1738–1822) and Charles Messier (1730–1817), amongst others, had the effect of expanding the scale and complexity of the then known universe in terms of time, distance and complexity. In addition to resizing the cosmos and extending its age, Herschel's work on proper motion had also demonstrated the probable non-centrality of the solar system. Taken together, these findings indicated to most astronomers that the universe contains millions (if not billions) of stars, many of which may have accompanying planetary systems, and that the place occupied by the planet earth is not necessarily of any greater significance than any other (presumed) planet.[4] These tentative conclusions were examined by theologians in general and proponents of natural theology in particular and found to justify a number of metaphysical viewpoints which accorded with their pre-existing religious assumptions.

The two key arguments *in extenso* applied to the abundance of stellar systems were those of plenitude and analogy. The argument from analogy suggested that other solar systems would resemble our own and the principle of plenitude suggested that God would not have bothered to create these additional worlds if He had not intended to extensively populate them, otherwise His creative energies might be said to have been devoid of purpose.[5] And furthermore, it was a reasonable expectation, at least for some students of religion and astronomy including Scottish theologian

[3] Michael J. Crowe, *The Extraterrestrial Life Debate 1750–1900: The Idea of a Plurality of Worlds from Kant to Lowell* (Mineola NY: Dover Publications, 1999), 3–5. For a general introduction to this topic, see also: Steven J. Dick, *Plurality of Worlds: The Origins of the Extraterrestrial Life Debate from Democritus to Kant* (Cambridge University Press, 1982) and Karl S. Guthke (tr. Helen Atkins), *The Last Frontier: Imagining Other Worlds, from the Copernican Revolution to Modern Science Fiction* (Cornell University Press, 1990).

[4] See George Gilfillan, *A Gallery of Literary Portraits* (Edinburgh: William Tait, 1845), 121, for a near contemporary reaction to the sense of great age and scale which arose from the work of William Herschel and others.

[5] Steven J. Dick, 'The Origins of the Extraterrestrial Life Debate and its Relation to the Scientific Revolution', *Journal of the History of Ideas*, 41, 1 (Jan–Mar., 1980): 3–27. William C. Heffernan, 'The Singularity of our Inhabited World: William Whewell and A. R. Wallace in Dissent', *Journal of the History of Ideas*, 39, 1 (Jan.–Mar., 1978): 81–100 at 84–86.

Thomas Chalmers (1780–1847), that these numerous other beings could and would worship Him:

> But there is no end of conjecture; and to the men of other times we leave the full assurance of what we can assert with the highest probability, that yon planetary orbs are so many worlds, that they teem with life, and that the mighty Being who presides in high authority over this scene of grandeur and astonishment, has there planted the worshippers of His glory.
>
> Did the discoveries of science stop here, we have enough to justify the exclamation of the Psalmist, "What is man, that thou art mindful of him; or the son of man, that thou shouldest deign to visit him?"[6]

This confidence in cosmic plenitude arose almost as an automatic corollary of natural theology. If one believes that the universe is a product of divine design, then the additional habitats which it was assumed to contain must also be populated, having been created and shaped by the same processes as our home planet. Although Chalmers asserted in this 1817 passage that God has a vast number of worshippers spread across "so many worlds", there is also at least a hint of a doubt, subsequently shared by Cambridge polymath William Whewell (1794–1866) amongst others, about the capability of the deity to take an interest in, and make propitiation for, every individual human being. The Psalmist clearly assumed a positive response to this question. Chalmers similarly employed the biblical text as a rhetorical device to be answered in the affirmative with an assurance of God's theistic care and attention.

Two broadly opposing nineteenth-century responses to the possibility of pluralist populations which reached peaks of adoption at different points in time can be identified. The earlier version was originally proposed by some of the religious commentators who opined on the topic at around the turn of the century. This species of cosmic theology sought to demonstrate the uniqueness of Man and to preserve a special revelation which only applied to the planet Earth. Other commentators, particularly those who, like Chalmers, had adopted an Evangelical and proselytising approach to religion in general, preferred to believe in a universe which was densely populated, containing many actual or potential worshippers of an omnicompetent deity.

[6] Thomas Chalmers, *A Series of Discourses on the Christian Revelation Viewed in Connection with the Modern Astronomy* (Glasgow: John Smith and Son, 1817), 32.

Thomas Paine and the *Age of Reason*

A significant challenge to revealed religion, which depended in part on the concept of pluralism, occurred towards the end of the eighteenth century when the potential superfluity of worlds and beings came to the attention of the political radical Thomas Paine (1737–1809). He employed the idea of cosmic plenitude, which was subsequently to be such an inspiration for Chalmers, as a significant plank in a tract entitled *The Age of Reason*. This work, based to some extent on his own interest in astronomy, argued against the literal truth of many elements of Christianity and thereby sought to reduce the (in his view) malign influence of the established church with its strong links to the ruling classes.

Historians remain divided on the question of to what extent Paine was in fact either a "filthy little atheist", i.e. a deist or whether, as now seems more likely to some scholars, he was pursuing a theistic, possibly Quaker, line of argument which sought to rescue the most important moral teachings of traditional religion from contemporary attacks.[7] Paine certainly came from a dissenting background, having been educated at a Quaker school where Latin and the Classics were eschewed, leading to the comparatively plain writing style of his adult work. In his role as the editor of a magazine which featured such political themes as the emancipation of slaves and the rights of women, he spent time with both the quietist 'Old Light' and more radical 'New Light' factions within the population of Pennsylvania, a city founded by and named for the Quaker William Penn (1644–1718). It has been claimed that these contacts, taken alongside the principles of Quakerism as a whole, were significant influences in his popular political polemics, including *Common Sense* (1776) and *The Rights of Man* (1791, 1792).[8]

Paine had also read at least some of the work of Baruch Spinoza (1632–1677) and would therefore have been exposed to the Dutch philosopher's complex concept of God's relationship with the material universe which had some affinity with pantheism and panentheism. Whether or not he accepted Spinoza's somewhat opaque metaphysics, Paine would

[7] Roy Porter, *Enlightenment: Britain and the Creation of the Modern World* (London: Allen Lane, 2000), 452–453.

[8] William C. Kashatus III, 'Thomas Paine: A Quaker Revolutionary', *Quaker History*, 73, 2 (Fall 1984): 38–61 at 41, 50–61. David C. Hoffman, '"The Creation We Behold": Thomas Paine's "The Age of Reason" and the Tradition of Physico-Theology', *Proceedings of the American Philosophical Society*, 157, 3 (SEPTEMBER 2013): 281–303 at 292–294.

certainly have concurred with his idea, which does indeed have overtones of Quaker philosophy, that priests should be stripped of their authority in order to allow for the development of human potential through the free expression of moral reasoning.[9]

He began writing his controversial attack on the credibility of Christianity in 1793 "just before he was imprisoned for ten months by the French revolutionaries, whom he had supported in *Rights*".[10] Perhaps the most remarkable aspect of this tract is that it was in part written at a time when Paine was incarcerated in the Luxembourg prison in Paris and "expecting to die hourly". He narrowly escaped the guillotine by means of a fortunately (or perhaps deliberately) misplaced chalk mark on the door of his cell. A fellow prisoner who was one of two physicians who attended Paine during a serious illness whilst in prison reported that he was "expecting not to see Paine alive in the morning, he [Paine] always expressed his firm belief in the principles of the book, and begged Mr. Bond should tell the world such were his dying sentiments". Paine's reported comment to Bond indicates that he was keen to secure his legacy as a radical polemicist in matters of religion as well as politics at a time when his prospects for survival looked poor and he appeared to be in imminent danger of both earthly and eternal retribution.

This account of Paine's imprisonment has to be read alongside several stoical deaths described in a hagiographic biography written by Moncure Daniel Conway (1832–1907) almost a hundred years later. It included, for example, Paine's report of the final words of Georges Jacques Danton (1759–1794): "'They will send me to the scaffold; very well, my friends, I shall go gaily.' Even so did Danton meet his doom."[11] These 'dying bravely' narratives seem curiously Christ-like, given the biographer's overall theme of Paine's success as a rationalist disposer of received religion.

Although Conway was of the opinion that Paine's thinking was strongly influenced by his Quaker background, other writers have positioned the style and content of *Reason* as having all the hallmarks of preceding works by eighteenth-century deists and that he "incorporated the themes and

[9] Maurice Bisheff, 'Paine, Spinoza, and the Radical Enlightenment', *Bulletin of Thomas Paine Friends*, 13, 2, Summer 2012: 7–8. Paine mentions reading the work of Spinoza on page 89 of Thomas Paine, *The Age of Reason* (1794 repr. London: Freethought Publishing, 1880).

[10] Crowe, *Extraterrestrial*, 162–164.

[11] Moncure D. Conway, *The Life of Thomas Paine*, vol II (New York: G. P. Putnam's Sons, 1892), 128–132.

the poetics of the rhapsodic mode of physico-theology".[12] Despite this apparent switch to rationalist and deist views on the primacy of empirical science as compared to the teachings of the Bible, however, Paine elsewhere continued to pursue the Quaker principle of listening to "the Inner Light, which guides the individual in his search for religious and moral truth" and thus to exercise his "private judgement in religious matters". As William Kashatus has argued:

> his later espousal of Jeffersonian deism was conditioned by the similarity of that philosophy to Quaker thought: both placed a fundamental significance on the individual's right to private judgment in religious matters, a de-emphasis on the significance of the Scriptures in shaping religious conviction, opposition to the mediation of clergy between God and the individual and an emphasis on humanitarianism.[13]

In his attack on the established churches and the clergy, Paine was one of the earliest authors to argue that the existence of multiple populated planets could undermine faith in the parochial belief system of Christianity. Like contemporary natural theologians, he proceeded from an initial assumption that a beneficent "Creator" had put in place a "society of worlds" to "administer [...] instruction to man", even if only "by their motion":

> Our ideas, not only of the almightiness of the Creator, but of his wisdom and his beneficence, become enlarged in proportion as we contemplate the extent and the structure of the universe. The solitary idea of a solitary world rolling, or at rest, in the immense ocean of space, gives place to the cheerful idea of a society of worlds, so happily contrived as to administer, even by their motion, instruction to man.[14]

The idea of multiple populated worlds, however also gave rise to significant religious doubts from Paine's point of view. Not surprisingly, he did not devote a great deal of time or attention to questions about the origins or nature of these hypothesised extraterrestrial beings: they are present primarily as a rhetorical device with which to advance his argument. He evoked their existence to create a sense of awed uncertainty as he

[12] Hoffman, 'Creation', 282–283, 303.
[13] Kashatus, 'Paine', 61.
[14] Paine, *Reason*, 44.

contemplated the implications for the capabilities and span of control of a single governing entity. Why would the supposed ruler of the universe expend such care and attention (and great self-sacrifice in Trinitarian theology) on a single insignificant planet such as the earth? Can a supreme being truly multitask to the extent of undertaking the 'cure of souls' on millions of worlds whilst also personally engaging in one redemption at a time? How would the Almighty, or a component thereof, find the time, and perhaps more importantly the penitential sufficiency, to undergo "an endless succession of deaths"?

> From whence then could arise the solitary and strange conceit that the Almighty, who had millions of worlds equally dependent on his protection, should quit the care of all the rest, and come to die in our world, because, they say, one man and one woman had eaten an apple. [...] In this case, the person who is irreverently called the Son of God, and sometimes God himself, would have nothing else to do than to travel from world to world in an endless succession of deaths, with scarcely a momentary interval of life.[15]

In addition to using the recently expanded scale of the universe as an argument against some of the core precepts of Christianity, Paine was here exploring a line of thought which was to become prevalent when people contemplated the potential meaning of the plurality of worlds during the early part of the nineteenth century: Was there a single act of divine sacrifice which took place on our home planet or is there a requirement for this immolation to take place on every populated body in the universe? Can one be saved if one is unaware that one has been redeemed? Could these arrangements be seen to be consistent with the idea of a universal theistic and caring deity? Would the received discourses of revelation and redemption need to be repeated across myriads of societies? As David Hoffman put it:

> Where earlier deists argued that a truly benevolent God would not arbitrarily confine the possibility of salvation to those who happened to have become acquainted with Christian revelation, Paine implies that it would be astronomically unworthy of a benevolent God to arbitrarily confine possibility of salvation to a single planet.[16]

[15] Paine, *Reason*, 44–45.
[16] Hoffman, 'Creation', 303.

The temporal association between impending death and Paine's insistence on revealing his innermost thoughts on establishment religion, bolstered by his use of pluralism as a logical absurdity in the form which subsequently came to be known as the "astronomical objection", can hardly be coincidental.[17] Whilst experiencing significant psychotropic affect during his illness and imprisonment, he made a vigorous case against traditional theology. Paine can be seen as having administered an alternative and intentionally perverse sacrament in which he confessed to the truth as he then understood it, whilst committing himself to the riskier side of 'Pascal's Wager', well knowing that an unwelcome outcome might be very shortly visited on him.[18] He can also be seen as performing a form of defiant, even arrogant, stoicism which, according to Conway, was typical of his co-prisoners. This colouration may of course have as much to do with late Victorian constructions of masculinity as it does with the actual lived experience of an inmate in a Parisian prison at the close of the 'Age of Sensibility'.[19]

Assuming that Paine had retained significant elements of Quaker theology, it is likely that he viewed his personal suffering as being a potential sign, indeed a prerequisite, of personal salvation. According to Carole Spencer:

> Suffering becomes redemptive because it is given meaning and purpose by identification with Christ, and viewed positively for the sake of Christ. When Christ is truly incarnated in the individual, that individual bears the cross of Christ. Participation in God means participation in the Cross, and in the passion of Christ [...] The belief that holiness involves suffering, even requires it, is much stronger in Quakers than other Puritan non-conformists.[20]

[17] Chalmers, *Discourses*, v.

[18] Pascal's wager is a form of apologetic in which Blaise Pascal (1623–1662) argued that it is sensible to believe in the existence of God and act accordingly since the reward for being right is in effect infinitely good and the penalty for an unbelief which turns out to be incorrect is equally limitlessly bad. See: Blaise Pascal, *Pensées* (New York: E. P. Dutton, 1958), Section III: para 233, 79–84.

[19] Northrop Frye, 'Defining an Age of Sensibility', *ELH (English Literary History)*, 23, 2 (June 1956): 144–152.

[20] Carole D. Spencer, 'Holiness: The Quaker Way of Perfection', *Quaker History*, 93, 1 (Spring 2004) 123–147 at 133. See also: Patricia Jalland, Death in the Victorian Family (Oxford University Press, 1996), 19–21.

Supporters and Opponents of Paine's Thesis

One of the earliest rebuttals of Paine's thesis occupied most of the first part of a lengthy book by the Baptist minister Andrew Fuller (1754–1815), who in 1799 cited scriptural support, or perhaps more accurately the lack of biblical opposition, for the very high value which God places on the salvation of a single soul:

> It is no exaggeration to say that the salvation of one soul, according to the scriptural account of things, is of inconceivably greater moment than the temporal salvation of a nation, or of all the nations in the world, for ten thousand ages. [...] THERE IS NOTHING IN THE SCRIPTURE DOCTRINE OF REDEMPTION WHICH IS INCONSISTENT WITH THE MODERN OPINION OF THE MAGNITUDE OF CREATION.[21]

Perhaps the single act of redemption which had taken place on earth would, unlike the multiple events imagined by Paine, be sufficient to save all the inhabitants of the universe? This certainly seemed to have been Fuller's view, although his hesitant assertion was couched via the self-evidently weak argument that the care extended by "the whole camp of Israel" to one victim of leprosy during their exodus from Egypt should be read as an analogy for God's care of the human race:

> The leprosy of Miriam was an event that affected the whole camp of Israel, nor did they proceed on their journies till she was restored to her situation: and it is not unnatural to suppose, that something analogous to this would be the effect of the fall and recovery of man on the whole creation.[22]

Fuller was a follower of a radical strand within the Baptist church. According to historian Catherine Hall, they were "committed to a notion of a universal family of man which stretched across the oceans, linking West Indians, Africans, Chinese and Indians in the embrace of its mission. [...] This was, in its own way, a transnational world".[23] Thus, his specific

[21] Andrew Fuller, *The Gospel, Its Own Witness: Or the Holy Nature and Divine Harmony of the Christian Religion Contrasted with the Immorality and Absurdity of Deism* (New York: Isaac Collins, 1801), 189. First published in 1801 with a preface by the author dated 10 October 1799.

[22] Fuller, *Gospel*, 202.

[23] Catherine Hall, *Civilising Subjects: Metropole and Colony in the English Imagination 1830–1867* (Cambridge: Polity, 2002), 1–2.

brand of nonconformist theology was a significant factor in his concerns for the redemption of extraterrestrial and terrestrial souls. From Fuller's perspective, humanity is to be seen as a peculiarly recalcitrant species within a populous broader creation and thus in particular need of God's paternalistic care in a manner analogous to the proselytising which he and his colleagues were concurrently carrying out amongst apparently unredeemed colonial cultures around the world.

A number of other clerics joined the debate on the side of traditional religion and a special revelation including the Oxford historian Edward Nares (1762–1841) and the Methodist theologian Adam Clarke (1762?–1832). Nares, who was "not known for his conciseness", published in 1801 a substantial tome, running to over 400 pages, in which he defended the tenets of revealed religion and the redemptive powers of the "Son of God" against the threat of deism.[24]

On the other hand, Paine's usage of pluralism as a critique of Christian theology was in due course adopted and further exploited by the poet Percy Bysshe Shelley (1792–1822), who had explicitly stated in his 1811 pamphlet *The Necessity of Atheism* that "the mind cannot believe the existence of a creative God".[25] It seems probable that Shelley and Paine were the two key opponents whom Chalmers later included within the impersonal categories of "the Treatises of the Infidel" and "Freethinkers".[26] Shelley, like his predecessor, was inclined to see the recently discovered "indefinite immensity of the universe" as providing supporting evidence for his challenge to revealed religion. In the notes which he appended to the end of his first large poetic work, *Queen Mab*, in which he discussed the multiple stellar systems through which Mab's magical chariot can roam, Shelley stated his views on the dissonance between the findings of astronomy, the concept of pluralism and a belief in the key tenets of Trinitarian Christianity which he saw as being built on "the childish mummeries of the God of the Jews":

[24] Crowe, *Extraterrestrial*, 172–173.

[25] Percy B. Shelley, 'The Necessity of Atheism' in Percy B. Shelley, (foreword by Henry S. Salt), *Selected Prose Works of Shelley* (London: Watts, 1915), 1–15 at 5. This pamphlet, the writing of which led to the expulsion of Shelley and his roommate from Oxford, was first produced as one of the notes to his poem Queen Mab. See: Percy B. Shelley, *Queen Mab: A Philosophical Poem* (London, 1813), 171. It was later printed as a pamphlet for private circulation, and the unused copies were deposited at William Clark's bookshop.

[26] Chalmers, *Discourses*, 5–12.

> The plurality of worlds,—the indefinite immensity of the universe is a most awful subject of contemplation. [...] It is impossible to believe that the Spirit that pervades this infinite machine, begat a son upon the body of a Jewish woman; [...] All that miserable tale of the Devil, and Eve, and an Intercessor, with the childish mummeries of the God of the Jews, is irreconcileable with the knowledge of the stars.[27]

Michael Crowe has pointed out that Shelley's work appears to have been influenced by the astronomical lectures of Adam Walker (1731–1821) as well as the writings of Lucretius. Walker had argued that a heliocentric cosmic order and the changing scale of the universe cast significant doubt on such topics as the existence of the devil and the location of hell, thus undermining some principal pillars of traditional punitive theologies. Shelley's personal preference was similarly to remove the concept of "eternal punishments for disbelief". To this end, he quoted William Herschel's view that "the sun possesses a cool interior against such authors as Swinden, who had designated the sun as the location of hell". His distaste for eternal damnation was, of course, subsequently also a topic of concern for Charles Darwin (1809–1882), whose grandfather Erasmus Darwin (1731–1802) had been an additional source of inspiration for the pluralism of *Queen Mab*.[28]

It is also evident that, in common with a number of other Romantic writers, Shelley was a regular user of opium, usually in the form of laudanum, "at times of excessive physical and mental stress".[29] Despite public protestations of reluctance, he openly referred to his laudanum habit in private correspondence. There is also an intriguing report of an incident when, having been refused access by the Godwins to their daughter and his future wife Mary (1797–1851), he "burst into the house and ran up to Mary offering a bottle of laudanum, and declared, 'By this you can escape from tyranny.'"[30]

[27] Shelley, *Mab*, 126–127.

[28] Crowe, *Extraterrestrial*, 170–171. Crowe references the prior work of Carl Grabo which demonstrated the close connection between Queen Mab and Erasmus Darwin's 1791 poem The Botanic Garden. Shelley, 'Necessity', 4.

[29] Victoria Berridge, *Opium and the People: Opiate Use and Drug Control Policy in Nineteenth and Early Twentieth Century England* (London: Free Association Books, 1999), 56–57.

[30] Katherine Singer, 'Stoned Shelley: Revolutionary Tactics and Women Under the Influence', *Studies in Romanticism*, Winter 2009, 48, 4: 687–707 at 701.

In his published poems, references to drug use were somewhat more circumspect, but there is good reason to believe that the lines in *Prometheus Unbound* which evoked "oracular vapour" and the sleep-inducing "Nepenthe" produced by "folded Elysian flowers" represented respectively nitrous oxide and opium.[31] It would therefore be reasonable to suppose that Shelley's heterodox musings on the religious implications of pluralism could in part be attributed to the results of the 'opium dreams', whose general effects on the subject's perceptions of time, space and theology were discussed in the preceding chapter.

William Paley's Natural Theology

To set the pluralism debate in a wider context, however, it should be borne in mind that not every religious apologist of the early 1800s was deeply concerned by Paine's diatribe and the responses it engendered. The author of the classic mainstream text of natural theology, William Paley (1742–1805), a Cambridge alumnus and Church of England cleric, was, perhaps surprisingly, remarkably reticent on the topic of pluralism. Paley's popular book, *Natural Theology*, first published in 1802, employed the apparent evidence of design in the universe as an argument for the existence of a deity whilst mostly ignoring the idea of extraterrestrial life. This seminal treatise became required reading for Cambridge students for several decades after its publication.[32] *Natural Theology*, his final and most widely read work, represented Paley's defence against a perceived upsurge in Enlightenment rationalism, deism and atheism, represented in particular by the works of the Scottish philosopher David Hume (1711–1776).[33]

[31] Shelley, *Prometheus*, 77, 84. Nepenthe (νηπενθής) is an allusion to a "powerful anodyne" which induces forgetfulness and was assumed by the Greeks to have originated in Egypt. See: Homer, (tr. Eric V. Rieu), *The Odyssey* (London: Guild Publishing, 1987), 70, (Book 4, Line 221).

[32] William Paley, *Natural Theology or Evidences of the Existence and Attributes of the Deity Collected from the Appearances of Nature* (Philadelphia PA: John Morgan, 1802). Aileen Fyfe, 'The Reception of William Paley's "Natural Theology" in the University of Cambridge', *The British Journal for the History of Science*, 30, 3, Student Papers (Sep., 1997): 321–335.

[33] Hume's more sceptical writings on the topic of revealed religion include the discussion of miracles in David Hume, *An Enquiry Concerning the Human Understanding* (1777 repr. Oxford: Clarendon, 1894), 114–131, and the posthumously published discussion of natural theology in David Hume, *Dialogues Concerning Natural Religion* (1779 repr. in Dorothy Coleman (ed.), *Dialogues Concerning Natural Religion: And Other Writings* (Cambridge University Press, 2007), 1–102.

For a review of eighteenth-century deists and rationalist religion, see Porter, *Enlightenment*, 96–129.

Like Chalmers, he declined to mention the targets of his tract by name, perhaps for fear of lending unintended weight to their utterances, thereby potentially inciting readers to examine their works in more detail, referring instead rather vaguely "to any scepticism concerning [natural theology] with which the present times are charged".[34]

In an interesting parallel with Paine, his teleological thesis was written at a time when, according to "his friend and closest companion during his final years", George William Meadley (1774–1818):

> That truly eminent man was then engaged in finishing his Natural Theology; but the completion of that great undertaking was frequently interrupted by severe accessions of a painful disorder, under which he had long labored, and which has since proved fatal.[35]

The "painful disorder" in question first surfaced two years before the publication of his *magnum opus* and had been diagnosed as a disease of the kidneys, "a violent nephralgic complaint", which was understood by Paley and his friends to be a terminal illness.[36]

In a published performance of a stoic self, reminiscent of Paine's reported fortitude in the face of death, Paley claimed to have found some positive aspects which arose from his otherwise painful and debilitating condition. Speaking of the existence of occasional periods of relief from the pain that he experienced, he claimed: "It has the power of shedding a satisfaction over intervals of ease, which, I believe, few enjoyments exceed" and that "A man resting from a fit of the stone or gout, is, for the time, in possession of feelings which undisturbed health cannot impart".[37] In 'The Goodness of the Deity', the penultimate chapter of *Natural Theology*, Paley thus contrived to convince himself of a theodicy in which suffering could be seen as a God-given good which both provided the positive experience of intervals of alleviation from pain and acted as a means by which terminally ill patients could reconcile themselves with their forthcoming

[34] Paley, *Evidences*, iv.

[35] Dan L. LeMahieu, *The Mind of William Paley: A Philosopher and His Age* (University of Nebraska Press, 1976), 2. George W. Meadley, *Memoirs of William Paley, D.D.* (Sunderland: J. Graham, 1809), 141.

[36] Ibid., 135.

[37] Paley, *Evidences*, 367.

death. As noted previously in the case of Paine, this was a personal atonement theology which had been a long-standing tenet of Quakerism. Around the turn of the century, it had also become a salient concept for Evangelicals and Unitarians.[38]

In complete contrast to Paine, however, Paley chose to disguise any doubts about conventional theology which he may have had at a time when he was seeking to demonstrate the underlying truth of the existence of a creative, and hopefully theistic, deity as he contemplated his own mortality. He thus succeeded in providing a key text for subsequent natural theology which relied on a teleological, almost deist, preference for rationality and empiricism which he felt should be seen as the foundation stone on which his earlier Christian constructions had been erected. "Over the course of the century, *Natural Theology* was read as a work of gentlemanly natural theology, as a work which could be used in a formal or informal education in science, and as a work of Christian apologetic."[39]

It is therefore somewhat ironic that a book which was subsequently regarded as a foundational Trinitarian tract had been written by an author who may have been for much of his career a closet Unitarian. "These suspicions were not unnatural at a time when Unitarianism attracted a number of intellectuals who were unable to accept orthodox doctrine, men like [...] Paley's friend and biographer Meadley."[40] Although Meadley defended his friend against the suspicion of being a fellow traveller of those "who are usually called Socinians", he also reminded readers that Paley had vigorously defended Edmund Law (1703–1787), Bishop of Carlisle, the father of Paley's closest friend at Cambridge. Law, like a number of Paley's contemporaries at Cambridge, was overtly in favour of the suspension of the requirement to subscribe to the Thirty-Nine Articles of the Anglican church. A lengthy pamphlet upholding the position adopted by Law was Paley's first published work.[41] It was appended to Meadley's biography apparently as evidence against the charge of Socinianism, but might well be taken by the reader as adding to rather than subtracting fuel from the fire.

[38] Jalland, *Death*, 49–51.

[39] Aileen Fyfe, 'Publishing and the classics: Paley's Natural theology and the nineteenth-century scientific canon', *Studies in History and Philosophy of Science*, 33 (2002): 729–751. Meadley, *Memoirs*, 150–166.

[40] Martin L. Clarke, *Paley: Evidences for the Man* (SPCK: London, 1974), 112.

[41] Meadley, *Memoirs*, 165–166. LeMahieu, *Mind*, 12, 16–19.

Meadley's account also made it clear that Paley was of a very latitudinarian disposition, being prepared to tolerate many variations in religious belief, including Methodism and Roman Catholicism. "For it was not to the adherents of any particular communion, that Dr. Paley was inclined to grant some partial boon; he was the consistent advocate of a complete toleration of all dissenters."[42] Paley was overtly willing to countenance fellow worshippers whose religious views might be considerably different from those of the Church of England and rather closer to heterodox strains of religion, including Unitarianism:

> If a Christian can think it an intolerable thing- to worship one God through one mediator Jesus Christ in company with any such as differ from him in their notions about the metaphysical nature of Christ, or of the Holy Ghost, or the like; I am sorry for it.[43]

The uncertainty amongst his contemporaries about Paley's possible Socinianism can also be explained by his apparent late adoption of Evangelical views at the height of their popularity around the turn of the century and coinciding, perhaps predictably, with his terminal illness.[44] Being aware of his forthcoming death, he may well have sought, as had Paine, to secure his legacy for posterity. In Paley's case, however, the favourable reputation to be preserved was that of an orthodox theologian for which purpose it was important to minimise any suggestion of irregularity.

Whether or not he was actually of a Unitarian persuasion, it is evident that he was of the opinion that personal beliefs should be based on a form of *sola scriptura*. He believed that only what can be found in the New Testament can be regarded as definitive. "In examining any point of controversy [...] I would advise you, before you take up a book on either side

[42] Meadley, *Memoirs*, 159–163.

[43] William Paley, 'A Defence of the Considerations on the Propriety of Requiring a Subscription to the Articles of Faith', in Meadley, *Memoirs*, v–vii and 9–43, footnote on p. 40 (this pamphlet follows after page 216 of Meadley's book).

[44] Boyd Hilton, *The Age of Atonement: The Influence of Evangelicalism on Social and Economic Thought 1785–1865* (Oxford University Press, 2001), 179, traces this idea of a late 'conversion' to comments made by John Joseph Gurney (1788–1847), an evangelical Quaker, and the evidence from Paley's posthumous sermons as reported in John D. Yule, 'The impact of science on British religious thought in the second quarter of the nineteenth century', PhD thesis submitted to Cambridge University (1976), 186. See also Meadley, *Memoirs*, 165–166.

of the question, to read the New Testament from beginning to end with a view solely to that one subject." In Paley's view, and that of some of his Cambridge peers, the scriptures long predated, and were therefore silent on the contents of, a set of Articles composed in a religiously divisive context by a committee composed primarily on one particular strand of protestantism. "The Articles, they maintained, were drawn up by fallible men, and to subscribe to them caused division and mutual dislike between Protestants."[45]

This insistence on accepting only the evidence of one's senses, that is, that which can be read, seen or otherwise personally experienced, also applied to his approach to *Natural Theology* in general and astronomy in particular. Paley's overall view of astronomy in terms of its usefulness as evidence of design was ambiguous at best. It was, in his opinion, "not the best medium through which to prove the agency of an intelligent Creator". He understood that no method was then available which would allow close inspection of heavenly bodies. They appeared only as "bright points, luminous circles", and therefore little could be directly known about their potential complexity of form or the nature of their inhabitants: a question which Paine, in his anti-establishment polemic, completely and conveniently overlooked. Paley also argued that as humans we are unable to imagine what it would be like to be a member of an alien society. This deficiency keeps us in "ignorance, moreover, of the *sensitive* natures, by which other planets are inhabited, [which] necessarily keeps from us the knowledge of numberless utilities, relations, and subserviencies, which we perceive upon our own globe".[46]

His indifference to the power, but not the sublimity, of astronomy can also be taken to mean that he did not feel that it had yet provided the ammunition that he needed to address the astronomical objection promulgated by Paine and Shelley. Paley therefore did not attempt to tackle such then unanswerable questions as whether or not other solar systems were analogous to our own and whether they posed similar challenges in terms of the possibility of geological, biological and theological evolution. On the other hand, the abundance of local heavenly bodies and the utility

[45] Clarke, *Paley*, 17, 18. The first quotation is taken by Clarke from a "document dating from Paley's Cambridge days, which Edmund Paley [son of William] himself reproduces". See also: William Paley and Edmund Paley, *The Works of W. Paley, D.D. And an Account of the Life and Writings of the Author*. vol 1, (Longmans: London, 1938), cclviii–cclix.

[46] Paley, *Evidences*, 287–288.

of their having been conveniently arranged by God into suitably placed and stable orbits around a stationary sun led him to the anthropocentric and teleological conclusion that the whole system was thus provided with convenient sources of light, heat and visual stimulation for human beings as well as demonstrating the wisdom of the author of the invariant laws of physics.[47]

To a natural theologian, the challenging implications of the disorderly and potentially metastable universe which had emerged during the preceding half century seemed to be inconsistent with his vision of a stable and well-ordered creation. Thus, the state of play in the early nineteenth century was that the key text of natural theology, whilst accepting the teleological purposes of creation, was somewhat of a *tabula rasa* in terms of any positive contribution to religious faith which might be secured from contemplation of alien intelligences. Paley's vaguely negative view can be contrasted with the Evangelistic enthusiasm and expansive interstellar populations subsequently promoted by Chalmers and later expanded to almost unbelievable proportions by another Scottish cleric Thomas Dick (1774–1857).

In 1829, Dick published a volume entitled *Philosophy of a Future State* in which he pursued a combination of natural theology and pluralism to what he conceived to be their logical conclusions. He calculated that there are 80 million visible stars, and that since, by analogy with the solar system, each star should have 30 or more satellites, there could be 2.4 billion inhabited worlds in the visible universe.[48] By 1840, and in continuing pursuit of his objective of demonstrating the vast size of God's congregation, Dick again expanded his earlier population estimate to encompass other stellar systems and solid-bodied comets (which he had latterly decided were also populated) and adduced a grand total for the visible universe of $60{,}573 \times 10^{18}$ inhabitants.[49] This huge number, which provides a suitably expansive precursor and counterpoint to the minimalist arguments of William Whewell (1794–1866) and Richard Anthony Proctor (1837–1888), represented a nineteenth-century high water mark in terms of the size of the population of the cosmos and the extent of

[47] Paley, *Evidences*, 289–305.

[48] Thomas Dick, *The Philosophy of a Future State* (New York: G. & C. & H. Carvill, 1829), 213.

[49] Thomas Dick, *The Sidereal Heaven and Other subjects Connected with Astronomy* (New York: Harper & Brothers, 1840), 317–318, 359–364.

God's kingdom. This vast congregation of potential worshippers exacerbated the dilemma for the advocates of pluralism which had been posed by Paine: in a vast and populous universe, how could God have time to exhibit even the slightest fleeting moment of concern for a single human being?

Samuel Taylor Coleridge

If Paley was uncertain about the utility of the possible existence of extraterrestrials, other writers held distinctly negative views. Samuel Taylor Coleridge (1772–1834), for example, was, at least towards the end of his life, a notable exception to the general tendency to pluralistic enthusiasm in the first half of the century. As he reached the final year of his life, and had perhaps more than an inkling that the end was approaching, he expressed his considerable doubts about the idea of pluralism:

> I never could feel any force in the arguments for a plurality of worlds, in the common acceptation of that term. A lady once asked me—"What then could be the intention in creating so many great bodies, so apparently useless to us?" I said—I did not know, except perhaps to make dirt cheap. The vulgar inference is *in alio genere*. What in the eye of an intellectual and omnipotent Being is the whole sidereal system to the soul of one man for whom Christ died?[50]

Coleridge has for a long time, and on his own admittance, been strongly associated with the consumption of opium, particularly during his poetic peak years. He claimed that he had "fallen excusably (because unavoidably) into this habit of eating opium" due to medical necessity. His argument was strongly refuted by a dubious Thomas De Quincey (1785–1856): "[Coleridge] suffers himself to fall into a captivity to this same wicked opium, deadlier than was ever heard of, and under no coercion whatever that he has anywhere explained to us. A slave he was to this potent drug not less abject than Caliban to Prospero."[51]

[50] Samuel T. Coleridge, 'Table Talk' (22 February 1834) in W. G. T. Shedd (ed.), *The Complete Works of Samuel Taylor Coleridge*, vol VI, (New York: Harper and Brothers, 1884), 502–503.

[51] Thomas De Quincey, *Confessions of an English Opium-Eater, Volume one in a series of De Quincey's works* (Edinburgh: Adam and Charles Black, 1862), 3, 8. This version had been substantially revised by the author in 1856.

De Quincey's scepticism was perhaps well founded. As a young man Coleridge had associated with several 'Romantic scientists', including Thomas Beddoes (1760–1808), Humphry Davy (1778–1829) and other 'Brunonians' who actively sought novel sensations and "were unusual in regarding opium as a stimulant". He joined them in experiments with nitrous oxide, a gas which Robert Southey (1774–1843) lauded as "a pleasure for which language had no name".[52]

A more mature Coleridge was apparently able to curtail but not to eliminate his opium usage. According to Earle Griggs and Seymour Porter, he became very dependent on this substance which he felt was essential to his ability to write:

> While it is unlikely that he ever returned to the enormous doses of his earlier years, he did continue to take […] what he considered sufficient opium to keep him "tranquil and capable of literary labor" […] Yet from 1808 on he struggled to emancipate himself, his latest attempt being in the spring of 1832, only two years before his death, and he never ceased to deplore his weakness. […] Opium, indeed, wrecked his life, but it is important to add that he was a constant sufferer from ill health.[53]

Another way of looking at Coleridge's comments is that although his addiction had become a permanent feature of his later years, he was, like many other users, able to manage his dosage at a level which enabled him to remain functional whilst no longer seeing or using opium as an inspirational stimulant.

In parallel with his achievement of chemical stability, Coleridge by this time was also exhibiting a rather more stable and conservative *Weltmodell* in matters religious and political. He had eventually returned to the fold of the liberal wing of the Anglican church after a youthful flirtation with being a "Unitarian, if not a Deist" and an unsuccessful attempt to devise an egalitarian 'panisocratic' utopia with his friend Southey and other

[52] 'Brunonianism' was the name given to a medical theory "based on the excitability of the animal body" which had originated with Edinburgh physician John Brown (1735–1788). Christopher J. Lawrence, 'The power and the glory: Humphry Davy and Romanticism' in Andrew Cunningham and Nicholas Jardine (eds.), *Romanticism and the Sciences* (Cambridge University Press, 1990), 213–227 at 215–219.

[53] Earle L. Griggs and Seymour T. Porter, 'Samuel Taylor Coleridge and Opium', *Huntington Library Quarterly*, 17, 4 (Aug., 1954): 357–378 at 363–364.

would-be supporters, most of whom fell away at an early stage.[54] His 1829 tract *On the Constitution of the Church and State* was post facto seen as an important, if belated, commentary on the Catholic emancipation debate, influencing the work of many Victorian writers.[55]

Coleridge's negative comments on the likelihood of the existence of extraterrestrial populations foregrounded his concern with the relative weight which the deity would place on a single human soul and thus represented an external expression of his rediscovered Trinitarian leanings and his ongoing repeated bouts of illness for which his at-one-time-prodigious consumption of opium had provided some relief but had also led to almost complete dependency. For some natural theologians, it had been axiomatic that hypothesised extraterrestrials must be members, or at least potential members, of an expanding congregation of worshippers. Coleridge's views, on the other hand, having developed his own conception of a specifically Anglican theology, were aligned with the alternative idea of a single redemption for the whole universe. He had clearly moved away from the pluralist views of some of his fellow Romantic poets, preferring to side with those commentators who sought to demonstrate the uniqueness of Man and to preserve a special revelation which only applied to the planet earth. This point of view had the additional virtue of being consistent with scripture and did not require the imagination of additional divine missions which might be seen as analogies of the extensive tour of the Americas envisioned by Joseph Smith (1805–1844), the founder of Mormonism.[56]

Interstellar Imperialism

The idea of a single Incarnation was echoed and expanded in an 1853 essay by William Whewell which we shall examine in more detail in Chap. 4.[57] Whewell's book also contained some empirical evidence which provided support for a view which became fairly prevalent in the late nine-

[54] James D. Campbell, *Samuel Taylor Coleridge: A Narrative of the Events of His Life* (London: Macmillan, 1894), 33, 255–256. Rachel Hewitt, *A Revolution of Feeling: The Decade that Shaped the Modern Mind* (London: Granta, 2017), 176–178.

[55] Andrew Elfenbein, 'Samuel Taylor Coleridge, On the Constitution of the Church and State (1829)', *Victorian Review*, Spring 2009, 35, 1 (Spring 2009): 19–23.

[56] Joseph Smith, *The Book of Mormon* (Palmyra NY: E. B. Grandin, 1830).

[57] William Whewell, *Of the Plurality of Worlds: An Essay* (London: John W. Parker and Son, 1853).

teenth century: a perspective which, with the benefit of hindsight, historians have generically labelled 'scientific naturalism'. If the genesis of the universe is the work of a deistic creator, or no creator at all, then its current state must have arisen as a result of a set of secondary laws or even stochastic and uncontrolled evolutionary processes. This viewpoint implied the corollary that intelligent life forms might be quite rare entities. Such a dichotomy still exists in current debates about the prevalence or otherwise of extraterrestrial life. There is an ongoing exchange of arguments between those who take the view that life will have arisen nearly everywhere that suitable conditions exist for its sustenance and the supporters of the 'rare earth' hypothesis who feel that our galaxy, sun and planet are unusually or even uniquely positioned to act as cradles of life. Religion no longer plays an overt role in this debate, but one might nevertheless not have to dig too hard to identify underlying metaphysical and even pseudo-theological beliefs which inform these opposing points of view.[58]

Finally, there is another interesting but rarely discussed question which should be asked with respect to pluralism, and which is still present in discussions on this topic: Why did (and do) many people assume that life on other planets is more 'advanced' than the human race? This issue seems be particularly apposite in the context of the early nineteenth century as that was an era during which there was also considerable debate between the proponents of monogenetic and polygenetic accounts of human evolution.[59] The latter viewpoint overtly cast other races as belonging to different and inferior species. Were human beings, by extension, now to be considered as a lesser developed and perhaps uniquely sinful race when compared to aliens? It was a period of continuing imperialist expansion, so it might be argued by analogy that just as Europeans had 'discovered' new worlds, subject populations and emerging commercial opportunities, so they in turn might be subsumed or exploited by colonialists from other planets. There may well also have been an element of a hangover from

[58] For a discussion of more recent debates surrounding cosmic plenitude, see Steven J. Dick, 'Cosmic Evolution: History, Culture, and Human Destiny' in Steven J. Dick and Mark L. Lupisellsa (eds.), *Cosmos & Culture; Cultural Evolution in a Cosmic Context* (Washington DC: NASA, 2009). As an example of the 'rare earth' view of the probability of extraterrestrial life, see John Gribbin, *The Reason Why: The Miracle of Life on Earth* (London: Allen Lane, 2011).

[59] For an outline of the monogenesis vs. polygenesis debate, see, for example, Robert J. C. Young, *Colonial Desire: Hybridity in Theory, Culture and Race* (London: Routledge, 1995), 6–13.

earlier cosmologies in which the earth was surrounded by higher planes inhabited by angels and other superior beings who stood in the role of intercessors between Man and God.[60] Whatever, then, was the cause of this inferiority complex, it is difficult to find many enthusiasts for pluralism in the nineteenth century who did not assume the superiority of the extra-terrestrial other.

[60] The hierarchy of created entities has been known since classical and Neoplatonic times as scala naturae or 'The Great Chain of Being'. In the seventeenth century, natural philosophers began to recast the chain as being spread throughout our own and other solar systems. See Arthur O. Lovejoy, *The Great Chain of Being: A Study of the History of an Idea* (Harvard University Press, 2001), 24–66, 99–143.

CHAPTER 3

"In Yonder Hundred Million Spheres": Thomas Chalmers and His Conversion to Pluralism

The truth within my mind rehearse
That in a boundless universe
Is boundless better, boundless worse.

Think you this mould of hopes and fears
Could find no statelier than his peers
In yonder hundred million spheres.

Alfred Lord Tennyson, 'The Two Voices'[1]

During the last few months of 1809, a young Church of Scotland minister was confined to bed in his isolated rural home in Fife whilst suffering from what he had good reason to believe was a fatal illness. A gloomy prognosis seemed almost inevitable to him as he had recently lost several relatives and close friends, including a sister, a brother and a surrogate father to "the same fatal malady". His son-in-law, William Hanna (1808–1882), reported: "For days and weeks he gazed upon the death brought thus so near, with eye intent and solemnized."[2] It is hardly surprising, therefore, that in this period of grief, suffering and morbid

[1] Alfred Lord Tennyson, 'The Two Voices' in Karen Hodder (ed.), *The Works of Alfred Lord Tennyson* (Ware: Wordsworth Editions, 1994), 172–185 at 172–173.

[2] Stewart J. Brown, *Thomas Chalmers and the Godly Commonwealth in Scotland* (Oxford University Press, 1982), 53. William Hanna, *Memoirs of the Life and Writings of Thomas Chalmers D.D., LL. D.*, vol I, (Edinburgh: Sutherland and Knox, 1849), 150–152.

H. Carlton, *Cosmology and the Scientific Self in the Nineteenth Century*, https://doi.org/10.1007/978-3-031-05280-4_3

expectation his thoughts turned to a contemplation of what might lie beyond: "The undiscover'd country from whose bourn / No traveller returns".[3]

Following this near-fatal illness, the effects of which lingered on for almost two years, it became clear that Thomas Chalmers (1780–1847) had also undergone a significant shift in his attitude to, and enthusiasm for, revealed religion. As we shall discover shortly, his previous approach to his clerical career had not been particularly diligent. After the episode at Fincraigs, however, he pursued a far more enthusiastic approach to religion. He succeeded, by dint of his powerful delivery of popular sermons in drawing relatively large crowds to hear him preach in his local parish of Kilmany, and in due course he was elected to the position of pastor for a much larger congregation in central Glasgow. After his promotion, he used this expanded platform not only to proselytise his personal version of religion but also to evince a notable interest in the meaning of the existence of extraterrestrial life and the contentious question of the age of the earth. As a result of the publication of his *Astronomical Discourses* in 1817, he became one of the most influential supporters and promoters of the concept of pluralism as Christian apology, and his book was still being reprinted into the 1870s.[4]

Thomas Chalmers was born in 1780 and raised in Fife as the sixth child of a devout but increasingly financially straitened Presbyterian merchant. At a relatively early age, he exhibited a precocious skill in the art of oratory and expressed an intention to be ordained as a minister.[5] He subsequently, however, experienced significant doubts about some elements of Christian theology. In his years as a mathematics student at the University of St Andrews, he described having a "scepticism [...] which affected the foundations of all religion, whether natural or revealed". In due course, nonetheless, he was licensed as a preacher by the Church of Scotland, after a rather turbulent period of employment as a tutor which ended prematurely, and was intellectually persuaded, or so he claimed, to a firmer conviction in the overall veracity of the scriptures "by reading Bishop Butler's Analogy of Natural and Revealed Religion, &c".[6]

[3] William Shakespeare, 'Hamlet, Prince of Denmark', Act 3, Scene 1.

[4] Michael J. Crowe, *The Extraterrestrial Life Debate 1750–1900: The Idea of a Plurality of Worlds from Kant to Lowell* (Mineola NY: Dover Publications, 1999), 182–189.

[5] Hanna, *Memoirs*, vol I, 3–9.

[6] Ibid., vol I, 24–33, 145–146. On p. 146, Hanna is quoting the words of "Rev. Mr. Smith", who was a colleague and "amanuensis" of Chalmers during 1807–1810.

Another early religious influence which casts an interesting light on his future conversion experience and his concept of a deity omnipresent in time and space, an important component of his commentary on pluralism, was that of the American divine Jonathan Edwards (1703–1758). Duncan Campbell Macgregor (1858–1943) noted that Chalmers was "absorbed in the metaphysical idea of God which he had learned from Jonathan Edwards [...] a Being to Whom 'every *where* is here and every *when* is now,'—to whom past, present and future are eternally one".[7] Edwards was deeply interested in the phenomenon of religious conversion having been both a participant in, and an observer of, a religious revival in the town of Northampton, Massachusetts: a process which he described in forensic detail in *A Faithful Narrative of the Surprising Work of God in the Conversion of Many Hundred Souls in Northampton* (1737).

During his early years in the ministry, Chalmers nonetheless remained distrustful of enthusiasts for Christianity and pronounced himself distinctly doubtful about the way in which the central tenets of atonement and redemption were interpreted by orthodox Calvinism, preferring instead the more sceptical conclusions reached by "free and rational inquirers":

> In what particular manner the death of our Redeemer effected the remission of our sins, or rather, why that death was made a condition of this remission, seems to be an unrevealed point in the Scriptures. [...] The tenets of those whose gloomy and unenlarged minds are apt to imagine that the Author of Nature required the death of Jesus merely for the reparation of violated justice, are rejected by all free and rational inquirers.[8]

It is interesting to note here that Chalmers views at this time were not so far removed from those of Thomas Paine (1737–1809), who, as we saw earlier, had poured scorn on the concepts of redemption and atonement in a pluralist universe.

Chalmers was at that time aligned with the Moderate wing of the Church of Scotland, a grouping of Presbyterians who "emphasized the themes of order, rationality, and social progress". Stewart Brown, in his

[7] Duncan C. Macgregor, 'Some Unpublished Letters of Dr. Chalmers', in Henry v. d. B. Copeland (ed.), *The Wimbledon & Merton Annual* (Wimbledon: Edwin Trim, 1903): 107–118 at 109–110.

[8] Hanna, *Memoirs*, vol I, 147. Hanna states that this is an excerpt from a sermon given by Chalmers at Kilmany.

biography of Chalmers, noted his "antipathy to Calvinism and religious enthusiasm" as well as a lack of "commitment to his clerical profession" and a somewhat lacklustre approach to serving his parishioners:

> In his pursuit of university preferment and literary reputation, he exhibited the most irresponsible aspects of early nineteenth-century Scottish Moderatism. He used his parish for his own support while he pursued a career in mathematics or political economy. He wrote his sermons hastily, and frequently left his parish for weeks or even months at a time. He neglected household visitations, the religious instruction of the young, and the other responsibilities of a parish minister.[9]

Chalmers' Conversion Experience

Chalmers subsequently underwent a considerable shift in his personal theology, becoming a prominent member of the Popular party, which opposed the more relaxed and rationalist approach of the Moderates. George Gilfillan (1813–1878), a near contemporary Scottish poet, ascribed this change to the experience of writing a specific text:

> Partly through the circumstance of being requested to write on the "Evidences of Christianity," for the Edinburgh Encyclopaedia, his mind was deeply and permanently impressed with a sense of religion. He felt that his preaching had hitherto been a "sham." With characteristic determination, he altered it from its foundation. Ceasing to be an "Ape of Epictetus," he became, for the first time a preacher of Christ crucified.[10]

Thus, in Gilfillan's view, the cause of Chalmers' noticeably altered sense of theological vocation and his move from Stoicism and "sham" preaching to his being an advocate of "Christ crucified" was to be found in a commission which he had undertaken to work on the *Edinburgh Encyclopaedia*. It will be shown shortly that this was not the key driver of Chalmers' change, although the experience of developing arguments in favour of Presbyterian theology provided the fertile intellectual soil into which the seeds of his subsequent conversion were sown. We presume that Gilfillan found this explanation personally satisfying because it illustrated the didactic benefits of a morally uplifting conversion story wherein the experience

[9] Brown, *Chalmers*, 44–49.

[10] George Gilfillan, *A Gallery of Literary Portraits* (Edinburgh: William Tait, 1845), 121.

of working with a suitably 'improving' text led to an important and desirable change. This interpretation of Chalmers' story thus demonstrated the efficacy of the right kind of written word: a category in which Gilfillan no doubt located his own literary efforts.

It seems, however, that the principal cause of Chalmers' changed *Weltmodell*, which in turn triggered his unexpected change of religious intensity was the accumulated effect of the pain, anxiety, isolation and extensive contemplation of mortality which accompanied an almost-terminal bout of "consumption" which afflicted him from summer 1809 to early 1810.[11] Michael Crowe, in his well-researched and comprehensive survey of the extraterrestrial life debate, noted that this significant life event happened "at the time of the deaths of some persons close to Chalmers and a serious illness he himself suffered". Brown similarly acknowledges that this was an important, indeed perhaps the most important, driver of the conversion process, although he adduces as an additional cause a failed romance with "a spirited beauty" which she probably terminated because of his "outbursts of temper". Macgregor, on the other hand, asserted that the relationship may have fallen through because of Chalmers' "reputation for a kind of eccentricity".[12]

A lengthier account of Chalmers' Damascene conversion can be found in Hanna's report of this significant turning point in Chalmers' life. Hanna's narrative is clearly structured as a deathbed conversion, foreshadowed by reports of several preceding deaths and illnesses. "But death was now to be no stranger: already had he borne away two of the family [his brother George and his sister Barbara] in his cold embrace; and two of his sisters were at this time threatened with the same fatal malady." Hanna particularly noted the stylistically pious and exemplary death of his neighbour and uncle, Mr Ballardie, who "had been a kind of second father" to him and who had "passed into eternity in a moment":

> She found him kneeling on a chair in the very attitude of prayer, but the spirit had fled—apparently without pain or struggle it had taken its departure; and from this lowly posture before the throne of grace on earth had passed into the presence of the throne of glory in the heavens. [...] It

[11] A German word meaning 'world picture' which has previously been employed to describe both the cosmological models of physical scientists and the creation stories found in traditional religions. It thus encompasses the two key epistemologies under discussion here.

[12] Crowe, *Extraterrestrial*, 183. Brown, *Chalmers*, 52–55. Also see endnote 20 on page 386. Macgregor, 'Unpublished', 110.

> seemed as if, once begun, the quick succession was to go on unbroken. A panic seized the family, as if one after another they were doomed to fall. Partaking fully of that panic, Mr. Chalmers believed that he was about to die.[13]

It is clearly relevant that Chalmers' illness followed so closely on the death of several close family and friends and so his condition must have been compounded with feelings of loss and grief as well as a sense of panic arising from a reasonably held conviction that he himself was about to die. Such a fate was not necessarily to be seen as unwelcome; however, if it could be recast, as in the case of Ballardie, as a transition to a higher and more desirable state of being. There were, as noted in Chap. 1, conventional expectations of a what a 'good death' should look like, and we find these echoed in a letter written by Chalmers in 1823 following the death of a sister of the woman who had earlier attracted Chalmers' unrequited romantic interest:

> And what an alleviation to your present sorrow, that you are not called upon to sorrow as others that have no hope, that on that death-bed which you have so recently witnessed there was a brightness and a glory that softened all its agonies, that amid the cruel sufferings of the flesh there was a spirit that bare up her on whom God was pleased to lay the hand of a refiner, and that in her case death was disarmed of its terrors and its sting, and she was more than conqueror through Him who loved her.[14]

One witness of Chalmers' illness was a Professor Duncan, who, in passing, provided evidence of both the "application of the very strongest medicines" and of the dramatic changes in the patient's appearance, including a very evident loss of weight and vigour:

> His illness, which was an affection of the liver, was such as to require the application of the very strongest medicines. "I visited him," says Professor Duncan, "at Fincraigs where he was under the medical treatment of Dr. Ramsay of Dundee, and I certainly never saw any person so much altered in the same space of time, being then greatly attenuated, while formerly he was corpulent. He was scarcely able to walk across the room. [...] It was a year or two before he recovered, and during that period he had much the appearance of an old man, of one who would never be able again for much exertion."[15]

[13] Hanna, *Memoirs*, vol I, 150–152.

[14] Chalmers to Mrs. Watson, 9 July 1823, Courtesy of the University of St Andrews Libraries and Museums, Special Collections, msdep93.

[15] Hanna, *Memoirs*, vol I, 151.

It is evident that the cause of Chalmers' malady was not well understood by "Dr. Ramsay", or any other observer, at that time. He may well have been suffering from tuberculosis as, presumably, had George and Barbara who succumbed to "consumption" in 1806 and 1808, respectively. We might accept the conclusion offered by his son-in-law that the "same fatal malady" also afflicted Thomas, even though his symptoms did not necessarily correspond to the contemporary "mythology of consumption" which defined the expected attributes of this disease. Thus, we note Hanna's parallel and conflicting allusion, perhaps derived from comments made by Ramsay, to "an affection of the liver". The latter was a portmanteau diagnosis which would be understood by contemporary medical professionals to cover a wide variety of possible causes and effects including a sympathetic reaction to disorders of the alimentary canal or the effects of heat and cold.[16]

We also don't know the exact nature of the medication prescribed for Chalmers' treatment, but we do know that during the early nineteenth century opium was very widely utilised as a sedative and palliative. "It would be almost easier to list those areas where it was never employed than to attempt to deal with every therapeutic possibility."[17] There thus exists a significant possibility that a psychogenic substance acted as an additional stimulus which coloured Chalmers' reveries and contemplations of eternity during this long period of illness and convalescence: "For four months he never left his room."[18]

Hanna's conversion narrative continued in the form of a reported interior dialogue which was intended to represent the thoughts which had exercised Chalmers during this isolation. In the course of these meditations, he had apparently reflected on "the feverish dream" of his past life, when "Blinded by the fascination of the things seen and temporal, he had neglected the things unseen and eternal". According to his son-in-law, he

[16] Basil Meyer, 'Till Death Do Us Part: The Consumptive Victorian Heroine in Popular Romantic Fiction', *The Journal of Popular Culture*, 37, 2, 2003: 287–308 at 289–290. Brown, *Chalmers*, 53, clearly asserts that the cause of Chalmers' illness was consumption. Hanna, *Memoirs*, vol I, 99–100, 152. Charles J. B. Williams, 'Affections of the Liver – Hepatitis', *The Boston Medical Surgery Journal*, 1845, 32: 149–155. https://doi.org/10.1056/NEJM184503260320801.

[17] Victoria Berridge, *Opium and the People: Opiate Use and Drug Control Policy in Nineteenth and Early Twentieth Century England* (London: Free Association Books, 1999), 66.

[18] Hanna, *Memoirs*, vol I, 151.

then resolved that "he would dedicate himself to the service of God, and live with the high aim and purpose of one who was in training for eternity".[19] A reader who is not closely wedded to Hanna's hagiographic objectives might interpret these remarks as meaning that Chalmers was aware that he had not lived up to the expectation of his 'calling' and that he was looking a little bit lightweight on the plus side of the ledger if he were to be called to eternal account in the not-too-distant future.

It is also hardly surprising that Thomas Chalmers' thoughts during this period turned to the "insignificance of time". His reported stream of consciousness included some reflections on the insubstantiality and transience of human existence when compared with "the magnitude of eternity" and thus on the potential futility of a life which perhaps has no continuation into an afterlife. As he wrote to his friend and colleague "the Rev. Mr. Carstairs", who lived in nearby Anstruther:

> My confinement [...] has fixed on my heart a very strong impression of the insignificance of time—an impression which I trust will not abandon me though I again reach the heyday of health and vigour. This should be the first step to another impression still more salutary—the magnitude of eternity. Strip human life of its connexion with a higher scene of existence, and it is the illusion of an instant, an unmeaning farce, a series of visions and projects, and convulsive efforts, which terminate in nothing.[20]

The Influence of Blaise Pascal

The changed perspectives which Chalmers adopted after his close call with death also appear to have been particularly influenced by his reading of *Pensées sur la Religion*, a rambling set of notes, published posthumously, in which Blaise Pascal (1623–1662) attempted to expound his personal theology and the *post facto* rationalisation of a form of conversion experience which lay behind a similar change of avocation from his explorations of mathematical theories to the promotion of a Jansenist strain of Christianity.

Pascal's journey towards Jansenism had been prompted by a near-fatal carriage accident which was followed by a long period of recovery and a series of doom-laden visions. "He swooned away, and was only restored with difficulty, and his nerves were so shattered that long afterwards,

[19] Ibid., 153–154.

[20] Chalmers to Rev Carstairs, 19 February 1809, quoted in Hanna, *Memoirs*, vol I, 152.

during sleepless nights and during moments of weakness, he seemed to see a precipice at his bedside, over which he was on the point of falling."[21]

Chalmers expressed his admiration for his predecessor's exemplary change of career direction and what would turn out to be an almost-parallel turn to a theology which was not too far removed from Calvinism:

> a man of the richest talents and whose youth was signalized by his profound and original speculations in mathematical science, but who could stop short in the brilliant career of discovery, who could resign all the splendours of literary reputation, who could renounce without a sigh all the distinctions which are conferred upon genius, and resolve to devote every talent and every hour to the defence and illustration of the Gospel. This, my dear sir, is superior to all Greek and to all Roman fame.[22]

One of Pascal's key arguments, which must have struck a chord with Chalmers, is that the process of conversion is not a purely intellectual exercise of the kind which might be achieved by a reading of Butler's *Analogy* or the contents of Brewster's *Encyclopaedia*, but is also dependent on an act of faith which defies or supersedes rational thought processes. "The heart has its reasons, which reason does not know [...] It is the heart which experiences God, and not the reason. This, then, is faith."[23]

Pascal's endeavour to "justify Christian ascent for reasons other than scientific" was similarly a source of inspiration much later in the century for George John Romanes (1848–1894) at a time when the Darwinian disciple was contemplating the roots of religious faith whilst struggling with the after-effects of a stroke:

> Several other theological books were read, and of all these the one which bears marks of most careful study is Pascal's 'Pensées.' [...]. It is the last book he read to himself in his own careful and studentlike fashion.[24]

[21] Norman Wilde, 'Scepticism and Faith in the Philosophy of Pascal', *The Harvard Theological Review*, 9, 1 (January 1916): 56–83 at 64–65, 70–72. The quotation is said to be from Jacques Benigne Bossuet's work but, unfortunately, Wilde does not give a precise reference to the source. Jansenism had strong similarities to Evangelical Calvinism including redemptionist soteriology, justification by faith and predestination.

[22] Hanna, *Memoirs*, vol I, 152. The entire quotation is taken from a further section of the letter to Mr Carstairs.

[23] Blaise Pascal, *Pensées* (New York: E. P. Dutton, 1958), Section IV, paras 277–278, 95.

[24] Robert J. Richards, *Darwin and the Emergence of Evolutionary Theories of Mind and Behavior* (University of Chicago Press, 1987), 372–373. Ethel Romanes, *Life and Letters of George John Romanes* (London: Longmans, Green, 1896), 342.

Romanes' thoughts on his reading material were relayed in a letter to his friend Francis Pagett (1851–1911), the dean of Christ Church College, Oxford, in which he asserted with Pascal that logical thought is not "the only means of research in regions transcendental".[25] Despite his acceptance of arguments based on faith rather than rationality as the basis for "a purified agnosticism", Romanes, unlike Chalmers, struggled to achieve a belief even in "a humanized Christianity (shorn of the troubling doctrine of the Incarnation)". Shortly before his death, however, he finally "did take Pascal's wager [...] he received Holy Communion", but as Robert Richards points out, his minimal version of religion avoided a potential conflict with his existing *Weltmodell*. It "was precisely the kind of spare faith that could be harmonized with the reign of natural law and evolutionary processes".[26]

Following what can be retrospectively seen as a seminal period in his life, Chalmers, inspired both by Pascal's ultimately successful bid for an alternative legacy and his own brush with death, took a much more active role in the life of the church. In 1814, he was adopted as the Evangelical candidate for the important role of minister of the Tron church in Glasgow, to which post he was in due course elected. During the selection process, his enthusiastic and powerful preaching overcame the concerns of some elders who had expressed their doubts as to whether or not his conversion was authentic. Their caution probably reflects their memories of his earlier diffidence and irregularity in matters of religion and represents further evidence of the suddenness of his change of direction as perceived by his contemporaries. They may also have entertained some genuine doubts about the sincerity of his espoused new position. He was indubitably motivated to disseminate Evangelical Christianity, but whether or not his particular soteriology cleaved closely to Presbyterian orthodoxy remained somewhat of an open question.[27]

Chalmers and the Age of the Earth

The reflections on eternity which had occupied him during his lengthy convalescence seem to have inspired an accompanying rethink by Chalmers on questions of ontology in general and the interpretation of Mosaic chronology in particular. This was a contentious topic as the assumed

[25] Ibid., 345.
[26] Richards, *Darwin*, 373–375.
[27] Brown, *Chalmers*, 91–92.

elapse of time since the creation had continued to increase, albeit in an uneven fashion, from the late eighteenth-century onwards based on arguments derived from astronomy as well as geology. As early as 1802, Frederick William Herschel (1738–1822), in a commentary delivered to the Royal Society, had proposed a minimum duration for the universe of around 1.9 million years.[28] The age of the earth itself, as opposed to the cosmos as a whole, had been significantly increased during the first decades of the nineteenth century by the bygone aeons which were adduced in ongoing debates about the meaning of the fossil record in which the thoughts of progressive creationists like the Reverend William Buckland (1784–1785) were contrasted with the more nuanced, less directional geological developmental processes of the kind proposed by Charles Lyell (1797–1875).[29]

On this question, post-conversion Chalmers, despite a continuing interest in mathematics and science which he had first exhibited during his education at St Andrews, nevertheless positioned himself as an upholder of Christian orthodoxy.[30] Shortly before his appointment as the minister of the Tron church, he became one of the earliest members of a group, identified by John Yule as "scriptural hermeneutists", who attempted to find room for geological ages within the Genesis account.[31] He was a prominent supporter of the idea (first mentioned in a sermon of 1814) that there is room for a considerable pause in activity between the creation of heaven and earth and the time when "the Spirit of God moved upon the waters" which he saw as the opening act of the traditional six days of creation (i.e. between the two sentences in the second verse of Genesis Chapter 1):

[28] William Herschel, 'Catalogue of 500 new nebulae', *Philosophical Transactions of the Royal Society*, 1802, 92.2: 477–528. Quoted in Stephen G. Brush, *The Nebulous Earth: The Origin of the Solar System and the Core of the Earth from Laplace to Jeffreys* (Cambridge University Press, 1996), 35.

[29] John H. Brooke, 'Science and the fortunes of Natural Theology: some historical perspectives', *Zygon*, 24, 1 (March 1989): 3–22 at 12.

[30] Brown, *Chalmers*, 5–6, 8.

[31] John D. Yule, 'The impact of science on British religious thought in the second quarter of the nineteenth century', PhD thesis submitted to Cambridge University (1976), 328, quoted in Boyd Hilton, *The Age of Atonement: The Influence of Evangelicalism on Social and Economic Thought 1785–1865* (Oxford University Press, 2001), 23.

> Between the initial act and the details of Genesis, the world for aught we know might have been the theatre of many revolutions, the traces of which geology may still investigate, and to which she in fact has confidently appealed as the vestiges of so many successive continents that have now passed away. The whole speculation has ministered a vain triumph to infidelity—seeing first that the Historical Evidence of Scripture is quite untouched by those pretended discoveries of natural science; and that, even should they turn out to be substantial discoveries, they do not come into collision with the narrative of Moses.[32]

This 'middle way' concept, which allowed that the "pretended discoveries of natural science" might in the end turn out to be "substantial", is known as the 'gap theory' of creation. It was not originally Chalmers' idea and was one of several attempts which emerged during the eighteenth and nineteenth centuries to harmonise the narrative of Genesis 1 with the findings of geologists and palaeontologists.[33] As Boyd Hilton has observed, Chalmers' version of gap theory provided a "theatre of many revolutions" which left room for several destructive and creative events of the kind which had been proposed by catastrophist geologists.[34] There was thus a deliberate ambiguity in his approach. He allowed himself and his audience to defuse the "vain triumph to infidelity" and to preserve a conceptual space into which he could fit a literal reading of Genesis, in line with traditional theology, whilst simultaneously providing an opportunity for potential new scientific discoveries to fill the gap.

Chalmers' tentative acceptance that deep time may well turn out to be a real phenomenon as opposed to the "allegorizing" proposed by scriptural geologists reflected his concern that no doubt should be cast on philosophical necessitarianism and the creator's capacity to control all elements of time and space which also explains his strong insistence on the original collocation of created matter. These beliefs were particularly important to a Calvinist because of the further implication which they carried that God was responsible for and had knowledge of the predestined fate of every soul.[35]

[32] Thomas Chalmers, *On Natural Theology*, Volume 1 (New York: Robert Carter & Brothers, 1857), 251.

[33] Ralph O'Connor, *The Earth on Show: Fossils and the Poetics of Popular Science, 1802–1856* (University of Chicago Press, 2007), 64–66, 213 n. 84.

[34] Hilton, *Atonement*, 150.

[35] David A. S. Fergusson, 'Predestination: A Scottish Perspective', *Scottish Journal of Theology*, 1993, 46, 4: 457–478 at 468–469.

Chalmers Addresses the 'Astronomical Objection'

Shortly after his arrival at his new post in Glasgow, and two years after the publication of the revolutionary and pluralist tract *Queen Mab* by Percy Bysshe Shelley (1792–1822), Chalmers chose to base his inaugural Thursday afternoon public sermon on the findings of 'modern astronomy'. In a series of lectures, Chalmers enthusiastically subscribed to the extraterrestrial life hypothesis. In direct contrast with Paine and Shelley, however, he saw the idea of a populous universe as a positive indication of the creative powers of a deity who operates in an expanded sphere in which "a higher scene of existence" could be located. Despite having to interact with an almost-uncountable number of souls, his omnipresent God was able to connect with and care for every being in such a way as to alleviate the anxieties evoked by his own recent brush with death:

> This body, which, upon the slightest derangement, would become the prey of death, or of woeful suffering, is now at ease, because he at this moment is warding off from me a thousand dangers, and upholding the thousand movements of its complex and delicate machinery.[36]

It is evident that Chalmers was at considerable pains both to alleviate his own intimations of mortality and to assuage the feelings of isolation and insignificance evoked in his audience as a result of the discoveries made by astronomers. He asserted that, despite the vastness of the universe, God nonetheless has sufficient powers to be able to take care of the humblest man, or even sparrow, within His overall creation:

> it is not for me, in the face of all this evidence, to reject it as a tale of imposture, because astronomers have told me that He has so many other worlds and other orders of beings to attend to [...] should a single sparrow fall to the ground without His appointment, then let science and sophistry try to cheat me of my comfort, as they may—I will not let go the anchor of my confidence in God—I will not be afraid, for I am of more value than many sparrows."[37]

[36] Thomas Chalmers, *A Series of Discourses on the Christian Revelation Viewed in Connection with the Modern Astronomy* (Glasgow: John Smith and Son, 1817), 107.
[37] Chalmers, *Discourses*, 110–111.

In the printed version of his sermons, first published in 1817, he maintained the rhythmic cadence of an originally oral presentation as he asked a series of rhetorical questions as a means of first building up and then bringing down a generic adversary whom he labelled "the Infidel":

> Is it likely, says the Infidel, that God would send his eternal Son, to die for the puny occupiers of so insignificant a province in the mighty field of his creation? Are we the befitting objects of so great and so signal an interposition? Does not the largeness of that field which astronomy lays open to the view of modern science, throw a suspicion over the truth of the gospel history? and how shall we reconcile the greatness of that wonderful movement which was made in heaven for the redemption of fallen man, with the comparative meanness and obscurity of our species?[38]

Chalmers knew that the "largeness of the field which astronomy lays open", that is, the scale and duration of the universe, was potentially a weakness as well as a strength of his argument. It could be used not only by natural theologians as evidence of God's creative powers but also by opponents of Trinitarian Christianity, like Paine and Shelley, who wished to "throw a suspicion over the truth of gospel history". The implied support in this first discourse for "the redemption of fallen man", which he further expounded in two succeeding discourses, also ran counter to his own pre-conversion scepticism as to the salvationary sufficiency of the Crucifixion. His solution to these rhetorical objections was to argue that the potentially strong case made by "the Infidel" should be countered by contemplation of both the omnipotence of, and the personal grace afforded by, "the same God who sits in the highest heaven, and reigns over the glories of the firmament, is at my right hand, to give me every breath which I draw, and every comfort which I enjoy".[39]

It is evident that Chalmers was seeking to demonstrate that his newly adopted religious position was compatible with his earlier educational experiences as well as building on the general level of interest in the sciences, particularly astronomy, in post-Enlightenment Scottish society to deliver a series of Evangelical sermons under the guise of objective natural philosophy. His choice of subject matter was considered both unusual and interesting. "The entire novelty of the discourse [...] excited the liveliest interest, not in his congregation alone, but throughout the whole

[38] Ibid., 54.
[39] Ibid., 53.

community."[40] His desired audience was not just the population of Glasgow in general but, according to Brown, the sermons were targeted in particular at the intellectual wing of the Moderate strand of Presbyterianism with which he had previously been affiliated:

> Chalmers's decision to focus upon the single theme of religion and astronomy reflected his continued interest in natural philosophy. However, it was also intended to demonstrate to the cultured despisers of Evangelicalism among the Glasgow social elite that his was not the religion of the unlettered fanatic or demagogue.[41]

Given this polemic objective for his course of lectures, the reception of both the oral and written versions must have exceeded his original ambitions. Of the various supporters of pluralism in the early to mid-nineteenth century, Chalmers was the most popular and widely read.[42] Despite being "an unattractive man" with a dishevelled appearance, Chalmers had an impressive oratorical delivery style, both when seated at his desk and in the imposing setting of the Tron church.[43] We learn from contemporary reports that the impact of his sermons included not only the "brilliant glow of a blazing eloquence", reported by the Baptist minister and essayist John Foster (1770–1843) on the basis of Chalmers' writing, but also a litany of leonine physical tropes described by Gilfillan:

> Excitement, convulsion, are fit yet feeble terms for his appearance, either at the desk or the pulpit. And yet, what painter has ever ventured to draw his preaching? And hence the dullness and paltriness of almost all the prints; they show the sibyl off the stool, the eye dim and meaningless, not shot with excitement, and glaring at vacancy; the lion sleeping, not the mane-shaking, tail-tossing, and sand-spurning lord of the desert.[44]

This dramatic delivery style was clearly a transient phenomenon which could be quickly replaced by the more contemplative mien which either he

[40] William Hanna, *Memoirs of the Life and Writings of Thomas Chalmers D.D., LL. D.*, vol II (Edinburgh: Sutherland and Knox, 1850), 87.

[41] Brown, *Chalmers*, 107.

[42] Ibid., 108. Brown cites several sources to demonstrate the popularity of Chalmers' sermons and the extraordinary success of the printed versions thereof.

[43] Ibid., 58.

[44] Hanna, *Memoirs*, 87. Hanna is quoting from a review of Chalmers, *Discourses*, by John Foster. Gilfillan, *Gallery*, 110.

or the artist deemed to be more appropriate when sitting for a portrait. A parallel account of Chalmers' delivery style has been pieced together by Brown from several contemporary sources:

> He might begin reading his sermons in a low voice, but as he progressed his pitch would heighten, his face would tauten, until with bulging eyes he hung over the pulpit, imploring each individual to recognize the power and accept the love of God. By this point the congregation forgot his ungainly figure and abandoned itself to his will. Men of all social orders felt his sincere passion for their souls. For his part Chalmers usually left the pulpit in a state of complete exhaustion, which often incapacitated him for days.[45]

The effectiveness of his post-conversion sermon-giving self was clearly far greater than his former somewhat lacklustre delivery. By choosing to use the performative style and theological language of a sermon, together with the mode of delivery from the authoritative elevation of his pulpit, Chalmers was selecting a template plucked from the praxis of the established church which would assist his delivery of a powerful and spiritually uplifting peroration in which he expended so much energy that he was left "in a state of complete exhaustion": a state which also recalls his preceding serious illness.

His appearances at the Tron church in Glasgow were attended by "congregated thousands" of Glasgow merchants, clerks and apprentices who, on the days when his sermons were to be given, left their desks early to attend Chalmers' popular performances. His sermons stimulated the imaginations and "human sympathies" of his audience by pointing out "the vastness and variety of those unnumbered worlds which lie scattered over the immeasurable fields of space". He thus provided his audience with a glimpse of the altered perspective on the scale of the universe and the insignificance of a single human lifetime which he had himself experienced:

> The spectacle which presented itself in the Trongate upon the day of delivery of each new astronomical discourse, was a most singular one. Long ere the bell began to toll, a stream of people might be seen pouring through the passage which led to the Tron Church. [...] heaven and its high economy

[45] Brown, *Chalmers*, 58. Also see endnote 36 on p. 386 for Brown's sources.

and its human sympathies and eternal interests, engrossed the mind at least and the fancy of congregated thousands.[46]

Chalmers perhaps also saw these congregations as a small sample of the numerous populations of the universe which either had been or needed to be redeemed by an experience of personal salvation. His post-conversion Evangelical theology helpfully promoted the idea that the salvationary offer arising from Christ's atonement was open to everyone, regardless of whether they yet had faith or not: an idea which extended to cover any extraterrestrial species who had not yet heard 'the Word'. It is certainly clear that he pursued an interest in terrestrial proselytising and had, according to theologian John Roxborogh, studied the "theories and practices of the Moravians, Baptists, the London Missionary Society, and the Church Missionary Society". As a result of "his enthusiastic, experimental approach", many of his own students became missionaries:

> His holistic philosophy, as well as his enthusiastic, experimental approach to mission, was reproduced and expanded by his students who became missionaries, educators, and church leaders around the world.[47]

He had begun his active support for missions to the unconverted populations of his home planet almost immediately after his serious illness. "The second wave of Scottish missionary enthusiasm commenced in 1809 [...] Chalmers's interest in the British and Foreign Bible Society was awakened shortly after his conversion."[48] "At the very time when Mr. Chalmers came to know the power and preciousness of the sacred volume, the enthusiasm on behalf of the British and Foreign Bible Society, which had been instituted a few years before, was at its height."[49] From that time onwards, and despite his partisan leadership of the Presbyterian breakaway church in the 1840s, he retained a broadly ecumenical sense of mission throughout his life: "his commitments to mission and principle were stronger than his commitment to any particular church, national or free."[50]

[46] Hanna, *Memoirs*, vol II, 88.

[47] John Roxborogh, 'The Legacy of Thomas Chalmers', *International Bulletin of Missionary Research*, (October 1999): 173–176.

[48] Brown, *Chalmers*, 62.

[49] Hanna, *Memoirs*, vol I, 263.

[50] Roxborogh, 'Legacy', 174.

Chalmers' Theology and Political Economics

Chalmers' post-conversion *Weltmodell* can be seen as having been constructed from a relatively consistent set of views which can be threaded together by taking a closer look at his modified theology. This was not as far removed from his earliest religious instincts as some of his peers may have assumed. In particular it is important to understand his slightly heterodox views on soteriology. It is often assumed that he was a conventional Calvinist who believed that a very limited population of the 'Elect' would or could be saved by the redemptive power of Jesus' self-sacrifice. In fact his personal views were somewhat more subtle than this. He held a sort of 'Catch 22' position in which he posited that it was impossible to know whether one was a member of the Elect and that it would be futile to even attempt the exercise as everyone is potentially due to be saved as a component of God's master plan. "In no place in the Bible is pardon offered to the elect, but to every person as one of the species." He asserted that "the bible if read with diligence and the Spirit given to your light upon the Bible if prayed for with sincerity" are the "means by which the poorest and humblest of men" might achieve Salvation. It was therefore not necessary or even appropriate for the individual to look for evidence of being one of the Elect since: "Those whom God hath ordained to eternal life he also ordains to the character that is meet for it."[51]

This involuntary form of predestination sits well with the philosophical necessitarianism which motivated some of the other elements of Chalmers' *Weltmodell*, including his belief in God's omnipotent control of time and space and a strong adherence to the idea of the original collocation of matter.[52] It also squares well with his acceptance that extraterrestrial beings may have been, or may in the future be, capable of salvation. Ironically, however, this avoidance of the need for evidence of having been 'chosen' would have downplayed the importance of Chalmers' conversion experience in his own mind, despite the subsequent attempt by Hanna to retrospectively foreground what he, for his part, perceived to be a very

[51] Samuel Dunn, 'Reminiscences of Dr. Chalmers', *The Wesleyan-Methodist magazine*, Nov. 1847, 3: 1077–1099. Chalmers to George Fortune, 5 October 1845, Courtesy of the University of St Andrews Libraries and Museums, Special Collections, msdep93. Thomas Chalmers, *Institutes of Theology*, vol II (Edinburgh: Sutherland and Knox, 1849), 384.

[52] The commonly held belief that God had set all existing matter in place at the time of creation and then set it in motion according to the laws propounded by Newton was clearly set in opposition to evolutionary cosmogonies.

important element in an account of his predecessor's life which eventually ran to four volumes.

Relatively few dramatic political changes can be observed in the narratives of the subjects of this book, presumably because the existential threats and emotional crises which they faced were not, as it happened, associated with situations in which a revised ideological stance could be expected to relieve their anxieties. Perhaps the clearest instance of a politically significant parallel paradigm change, however, was that of Thomas Chalmers, who, following his conversion to Evangelism, espoused a strong belief in a divinely ordered moral universe in which people would undertake 'right action' once they had been suitably educated.

A youthful Chalmers had been prone to occasional vituperation against class privilege: "in his anger with the gentry for ignoring his abilities and refusing to acknowledge his family's proper status in the hierarchy of social orders, his thought was beginning to move in a radical direction." Nonetheless, his anomie did not lead him to embrace the kind of revolutionary political theory adopted by many of his contemporaries; indeed, his thoughts were "more a function of thwarted ambition than any defined social ideal".[53] There does not appear to have been any period in which he took up a truly radical point of view which would have been more congruent with the egalitarian ideas of 'infidels' like Paine and Shelley or the socialist Robert Owen (1771–1858).[54]

As a young man, he had also expressed support for an agrarian model of political economy which he saw as a counter to Napoleon's an attempted blockade of Britain, known as the 'Continental System', and which would alleviate the prevalence of marginal subsistence in his native Fife.[55] Following his appointment to the Tron church, however, his personal experience of voluntary work amongst Glasgow's poorer citizens, combined with his reading of the works of Thomas Malthus (1766–1834), convinced him of the need to improve their lot by means of behavioural change rather than financial support. This led him to what might at first sight seem to be a perverse conclusion: that the provision of relief for the poor would fail to produce any improvement in their morality, and hence

[53] Brown, *Chalmers*, 19.

[54] Laura J. Snyder, *Reforming Philosophy: A Victorian Debate on Science and Society* (Chicago: University of Chicago Press, 2006), 293–294.

[55] The Continental System was an attempted blockade of British commerce and manufacturing industry during the Napoleonic wars and was partially successful in reducing exports to mainland Europe in the period 1802–1806. See: Brown, *Chalmers*, 35–42.

their health and welfare, but would instead have the opposite effect.[56] Chalmers, like his fellow Presbyterians and some, but not all, natural theologians saw poverty as a sign of moral failure and irreligion. He therefore supported the abolition of the Poor Laws because he believed that there was an opportunity to improve the "economic comfort" and "physical comfort" of his fellow Glaswegians by enhancing their "moral worth" and "prudence":

> so as to ensure an inseparable connection between the moral worth and the economic comfort of a people [...] political economy is but one grand exemplification of the alliance, which a god of righteousness hath established, between prudence on the one hand, and physical comfort on the other.[57]

On the other hand, Chalmers' apparently austere approach to the provision of relief should not be read as a lack of empathy or even generosity. He was quite successful at soliciting donations from his better-off followers which he then proceeded to discreetly channel to those whom he deemed worthy of support:

> I had, Sir, in common with thousands, often witnessed that thorough devotedness to the cause of the labouring classes which you manifested whilst dwelling among us—& had heard much of the active though often secret benevolence of which you then made our City the theatre—& had admired that eloquence of which the improvement of the workman was not seldom at once the subject and the aim.[58]

It seems, then, that there was a slow but significant change in Chalmers' views on politics and political economy following his theological conversion event. It had become his view that there was a very strong connection between religious observance and the well-being of the population. His gradual shift from advocacy for agrarian self-sufficiency towards a Malthusian approach to the urban poor seemed to be related to the cultural and economic context of his parishioners as well as to his choice of

[56] Ibid., 117–129.

[57] Chalmers, *On the Power*, 52–54.

[58] James A Begg to Thomas Chalmers, 23 July 1825, Courtesy of the University of St Andrews Libraries and Museums, Special Collections, ms30385, letter 81.

reading matter. Post-conversion Chalmers was, however, a quite complex character who could, at different times, appear to be either compassionate or dispassionate in his interactions with his fellow human beings.[59]

Extraterrestrial Evangelism

It is clear from contemporary sources that a relatively young Thomas Chalmers appeared to have undergone a significant conversion from relative indifference to an enthusiastic adoption of Calvinist orthodoxy as a result of personal loss and a near-fatal illness. He expressed his new-found Evangelism both in his exhortations from the pulpit and in the form of a book which sought to expand considerably the magnitude of God's congregation. The quantity of potential souls in the universe had expanded during the early nineteenth century as a result of new perceptions of the age and scale of the cosmos, and it was by no means clear that a single redemptive act would be sufficient to secure their salvation. His earlier meditations on the depth and subjectivity of time, however, provided a potential solution to this problem for Chalmers. In response to the arguments of "the Infidel who urges the astronomical objection to the truth of Christianity" that the deity could not possibly have had time to arrange penitential processes on the potential infinity of other worlds, Chalmers was able to retort that God, who can treat "a thousand years as one day", has "the mighty roll of innumerable ages" to work in.[60]

In the sphere of political economics, he perceived that the antidote to poverty and destitution lay in the doing of good works by the bourgeoisie and via the moral and religious improvement of the working classes. These solutions drew significant criticism from contemporary theologians and politicians, being seen as too "parochial and voluntarist" especially when he attempted to apply them in a large city.[61] Chalmers' reformed *Weltmodell* nonetheless prompted him to adopt a form of holism in which the intentions of the deity and the truth of Christianity could be assisted on earth (and presumably on other planets) by the actions of right-thinking and religiously motivated individuals.

Thomas Chalmers reacted to his personal brush with mortality, during which he reflected on his own prior indolence, by conceptualising the

[59] Brown, *Chalmers*, 283–284.
[60] Chalmers, *Discourses*, 155–157.
[61] Hilton, *Atonement*, 62.

Creator as a stern taskmaster who expected people to achieve their own salvation by means of sexual self-control and moral improvement.[62] The controlling cleric and his judgemental God represented a significant shift from an earlier 'Moderate' deity whose disengaged and indolent earthly servant had applied minimal effort in carrying out his parochial duties. This change became evident shortly after his lengthy convalescence when, according to Brown:

> Chalmers appeared upon the quiet, Moderate-dominated religious scene of northern Fife as an impassioned Evangelical preacher, proclaiming man's total depravity and alienation from God, and his absolute dependence upon divine grace for salvation. With powerful oratory, he now pleaded with men to embrace the Evangelical views which he had previously denounced.[63]

This paternalistic God was evidently compatible with the emergence of a remade and re-energised Chalmers, an autocratic leader who became the first Moderator of the Free Church following the great disruption of 1843:

> He employed the parallel hierarchies of Church Extension committees and Church courts to impose his personal will upon the Church and nation. In his personal life, he was often domineering and dictatorial. [...] He could be a tyrant at home with his family. He had an explosive temper, and was quick to perceive a personal affront, often when none was intended. In most of his relationships, he demanded absolute loyalty.[64]

Chalmers in due course became a highly influential theologian and church leader. His thoughts on pluralism, in particular, reached a wide audience and set the tone for subsequent debates on this topic. He is also a very evident example of the connection between severe illness, religious conversion and the stimulation of the imagination which the poet Thomas Stearns Eliot (1888–1965) also noted in his introduction to Pascal's *Pensées*:

[62] For evidence of Chalmers' moral paternalism, see: Hilton, *Atonement*, 88–89, 208. Hilton's view is summarised succinctly in Crosbie Smith, *The Science of Energy: A Cultural History of Energy Physics in Victorian Britain* (London: Athlone, 1998), 20.

[63] Brown, *Chalmers*, 44–50.

[64] Brown, *Chalmers*, 334, 284.

> We know quite well that he was at the time when he received his illumination from God in extremely poor health; but it is a commonplace that some forms of illness are extremely favourable, not only to religious illumination, but to artistic and literary composition. A piece of writing meditated, apparently without progress, for months or years, may suddenly take shape and word.[65]

It is also evident that there was a proselytising intention which connected Chalmers' attitude to the potentially convergent paradigms of Evangelism and the existence of extraterrestrial life. He was certainly of the opinion, expressed in his *Astronomical Discourses*, that the Incarnation and Atonement were key moments in an eternal universal struggle between the forces of good and evil and that this battle continues to take place in each and every human heart:

> To an infidel ear, all this carries the sound of something wild and visionary along with it. But though only known through the medium of revelation; after it is known, who can fail to recognize its harmony with the great lineaments of human experience? Who has not felt the workings of a rivalry within him, between the power of conscience and the power of temptation?

Having converted from relatively passionless moderation to a newly found enthusiasm for the core messages of his chosen form of Christianity he sought to spread the 'good news' to his fellow humans and thus to expand the congregation of God's worshippers in support of this cosmic struggle. In seeking to overcome "this presumption of the Infidel astronomy", he was at the same time assuring both himself and his readers of the infinitely multi-tasking nature of God's theistic care.[66]

[65] Thomas S. Eliot, 'Introduction' in Pascal, *Pensées*, vii–xix, at x.
[66] Chalmers, *Discourses*, 211–214.

CHAPTER 4

"What Is Man if Thou Art Mindful of Him?": William Whewell and the Unique Revelation

Tho' world on world in myriad myriads roll,
Round us each with different powers,
And other forms of life than ours,
What know we greater than the soul?

Alfred Lord Tennyson, 'Ode on the Death of the Duke of Wellington'[1]

Taking the thoughts of Thomas Paine (1737–1809) and Percy Bysshe Shelley (1792–1822) as examples from Chap. 2, we saw that the concept of interplanetary pluralism was not always deployed to promote traditional religious beliefs. The comments made by Samuel Taylor Coleridge (1772–1834) in 1834 similarly remind us that an opposing view was likewise not automatically a denial of conventional theology.[2] A unique revelation applicable only to man could potentially be seen to sit much better within a sparely populated universe: an argument which was to be subsequently promoted by William Whewell (1794–1866), one of the many people who had studied the *Astronomical Discourses*, written by pro-pluralist Thomas Chalmers (1780–1847), with interest and appreciation.

[1] Alfred Lord Tennyson, 'Ode on the Death of the Duke of Wellington' in Karen Hodder (ed.), *The Works of Alfred Lord Tennyson* (Ware: Wordsworth Editions, 1994), 400–407 at 406.

[2] Samuel T. Coleridge, 'Table Talk' (22 February 1834) in W. G. T. Shedd (ed.), *The Complete Works of Samuel Taylor Coleridge*, vol VI, (New York: Harper and Brothers, 1884), 502–503.

H. Carlton, *Cosmology and the Scientific Self in the Nineteenth Century*, https://doi.org/10.1007/978-3-031-05280-4_4

Whewell, unlike Chalmers, had had plenty to say on the probability of life on other planets prior to experiencing an ongoing personal crisis which accelerated his drive to reverse his earlier views and to conclude that the earth may be the only planet occupied by organic life forms. Current explanations of Whewell's change of mind are based on intellectual arguments which revolve around his version of metaphysics or his aversion to a specific text. Later in this chapter, however, a closer reading of his personal correspondence will be used to show that a significant affect which impinged on him at the time of his rejection of pluralism was the ultimately fatal illness from which his wife was suffering and which led to a shared and understandable desire to secure her eternal future.

Whewell was the son of a Cumbrian carpenter who had exhibited such academic promise at school in Lancaster and Heversham that he won a scholarship to attend the University of Cambridge.[3] As a young Fellow at Trinity, the college of which he subsequently became master, he had evidently followed a reading recommendation which he had received from his sister:

> I am glad you have been reading more of Chalmers's sermons; I have only read one or two of his astronomical ones, but I think they are all that you describe them.[4]

More evidence of Whewell's early interest in, and potential support for, pluralism came in the form of a sermon which Whewell preached at Cambridge, shortly after his ordination into the Church of England, in which he expressed his then firm belief in extraterrestrial "responsible beings" despite the challenge to universal theistic care which might arise from these "joint claimants of God's notice":

> That the earth which man inhabits is one among a multitude of worlds which roll on their appointed paths in the populous universe, at intervals which it exhausts the imagination to span [...] these are not the reveries of

[3] Isaac Todhunter, *William Whewell, D. D. Master of Trinity College, Cambridge: An Account of His Writings with Selections from His Literary and Scientific Correspondence*, vol 1 (London: Macmillan, 1876), 1–2.

[4] Whewell to Anne Whewell in Janet M. Stair Douglas, *The Life and Selections from the Correspondence of William Whewell, D. D. Late Master of Trinity College Cambridge* (London, 1881), 73–74. The letter is undated but was included by Stair Douglas with other correspondence from 1822.

> idle dreamers or busy contrivers;—but for the main part, truths collected by wise and patient men on evidence indisputable [...] founded upon analogies which most will allow to reach at least some degree of probability [...] Now it may be natural that a believer [...] upon thus learning as it were the population of the whole universe let in upon him as joint claimants of God's notice and objects of his care, feel, so to speak, alarmed and dispirited; as if it were too much to expect that his preservation and redemption and satisfaction should occupy the mind which is charged with the care of ten thousand worlds and systems.[5]

Whewell went on to make use of the same excerpt from Psalm 8:4 ("What is man that thou art mindful of him?") which Chalmers had deployed in his *Discourses*. The primary thrust of this sermon can be seen as having been derived with little deviation from Chalmers' original work. In reading this passage, we understand that Whewell in the 1820s was in general agreement with the pluralist arguments then very much in vogue amongst natural theologians, and though he shared with others some theological doubts and a concern that the underlying basis of Christianity and the idea of a personal theistic deity could be undermined by this position, he accepted that an expanded universal population could nonetheless be seen as a positive indication of the omnipotence of the creator. It is, however, at least a little ironic that, despite his later erudite ruminations on inductive philosophy and Baconian science, Whewell at this stage was quite happy to accept as "evidence indisputable" some fairly speculative ideas based on the argument from analogy to the effect that the universe is populated by numerous living and intelligent beings.

William Whewell's Declining Support for Pluralism

Whewell was invited to author a *Bridgewater Treatise* because of his previously displayed sympathy with the aims and arguments of natural theologians. It is unsurprising, given his interest in the work of Chalmers, that he chose *Astronomy and General Physics considered with Reference to Natural Theology* as the title and theme for his contribution to the series. Whewell dedicated one section of this book, first published in 1833, to a discussion of the argument from final causes in which he posited that considerable caution should be exercised by natural philosophers when investigating

[5] Sermon preached by Whewell on 18 February 1827 in Todhunter, *Whewell*, vol 1, 26.

the origins of objects and phenomena in the universe. Whewell, unlike Chalmers, advised philosophers "not to assume that we know the objects of the Creator's design". He advocated instead the study of the secondary laws which generate many of the aspects of creation now visible to man:

> Final causes are to be excluded from physical inquiry; that is, we are not to assume that we know the objects of the Creator's design, and put this assumed purpose in the place of a physical cause. We are not to think it a sufficient account of the clouds that they are for watering the earth, (to take Bacon's examples) or "that the solidness of the earth is for the station and mansion of living creatures." The physical philosopher has it for his business to trace clouds to the laws of evaporation and condensation; to determine the magnitude and mode of action of the forces of cohesion and crystallization by which the materials of the earth are made solid and firm.[6]

Whewell was arguing for a method of scientific enquiry which favoured a process of discovering proximate materialist explanations for the phenomena of nature rather than trying to second guess the mind of God. This will turn out to be one of several points of view held by Whewell in the 1830s which were to be modified in the early 1850s at a time when Whewell appeared to be much more confident that he understood the final cause of creation.

The treatment of pluralism in his *Bridgewater Treatise* was not expansive. Whewell devoted a few comments in a section entitled 'On the Vastness of the Universe' to a consideration of "these countless species of creatures" who inhabit the known and as-yet-unknown universe, concluding with an encomium to the powers of the "Maker and Master of all" which was fully in keeping with contemporary natural theology:

> On the contrary, those magnitudes and proportions which leave our powers of conception far behind [...] may reasonably serve, not to disturb, but to enlarge and elevate our conceptions of the Maker and Master of all; to feed an ever-growing admiration of His wonderful nature; and to excite a desire to be able to contemplate more steadily and conceive less inadequately the scheme of his government and the operation of his power.[7]

[6] William Whewell, *Astronomy and General Physics Considered with Reference to Natural Theology* (London: William Pickering, 1833), 264.

[7] Whewell, *Astronomy*, 212.

The effusive support for pluralism exhibited by Whewell in 1827, somewhat less energetically argued in 1833, can be clearly contrasted with the evident change of mind which reached its full flowering in *Of the Plurality of Worlds: An Essay*, published anonymously in 1853, and in which he sought to almost entirely refute pluralism. There are some indications that this change of mind began to take place in the late 1840s in the form of the following comments made by Cambridge mathematician and Whewell biographer Isaac Todhunter (1820–1884) with respect to a paper Whewell gave at that time:

> I have already said on page 188 that this memoir contains, in the form of a suggestion, an anticipation of the main position of the *Essay* on the Plurality of Worlds, namely, that man is the principal object of creation.[8]

The passage to which Todhunter referred was inserted as a worked example, rather than the main thesis, in a paper on metaphysics and methods. This discussion gives us an insight into Whewell's views on both pluralism and the nebular hypothesis in 1848. It should be particularly noted that his examples of extraterrestrial organisms were now restricted to "vegetable and animal life" with no mention of intelligent beings:

> the Idea, such as the motions of satellites, the relations of the elements of planets, the existence of vegetable and animal life upon their surfaces. But if all these Facts, when fully studied, should appear to be included in the general Idea of Nebular Condensation [...] we are also to recollect that the Facts may be compared with another Idea, no less large and striking [...]. The other Idea which I speak of is the Idea of Man as the principal Object in the Creation; to whose sustenance and development the other parts of the Universe are subservient as means to an end.[9]

The idea of the presence of living organisms on other planets was, in this article and at this point in Whewell's life, taken to be evidence against the nebular hypothesis. This argument was combined with other "Facts" which Whewell alluded to in order to cast doubt on nebular cosmogony. His evident opposition to nebular evolution in the late 1840s probably

[8] Todhunter, *Whewell*, vol 1, 317.

[9] William Whewell, 'Second Memoir on the Fundamental Antithesis of Philosophy', *Transactions of the Cambridge Philosophical Society*, vol VIII, part I (Cambridge: University Press, 1849), 614–662 at 615–616.

arose at least in part because it had been claimed as a proven theory by the author of the recently published evolutionary tract entitled *Vestiges of the Natural History of Creation*. Importantly, in his view, the stochastic and uncontrolled outcomes of the apparently random development processes described in this controversial monograph conflicted with "the Idea of Man as the principal Object in the Creation".[10]

The lines of argument which were presented as a worked example in Whewell's 1848 lecture are consistent with his frequent employment of both idealist metaphysics and inductive reasoning. His emerging support for the concept that reality is indistinguishable from a mental construct in the minds of an observer has been posited by some historians as an important driver for his subsequent changes of emphasis and belief in 1853. Whewell's attachment to idealism is usually attributed to his reading of Immanuel Kant (1724–1804) and of continental philosophy in general. This putative affiliation with the views of the 'German Romantics' offered potential dangers as well as benefits because "extreme idealist philosophy dissolved the distinctions between man, God and nature, and challenged the authority of Revelation".[11]

Richard Yeo has also identified a revival of seventeenth-century Neoplatonism in the work of Whewell and has, like Laura Snyder, pointed out that in addition to metaphysics, Whewell saw theology as a fundamental epistemology which was not to be divorced from the world of science:

> He proclaimed the religious and intellectual value of natural science while repudiating the empiricist philosophy of science promoted by the utilitarians [...] and offered an idealist philosophy of science which might satisfy the requirements of a Christian natural theology.[12]

By the late 1840s, as we have just seen, Whewell clearly shared with some supporters of idealism the concept that man occupied a special position in the schema of the cosmos. His principles included a belief, which he shared with Chalmers, that the ideas or laws which govern the universe are 'necessary' in the sense that they pre-existed in the mind of the deity. In his view, therefore, man is the sole entity, other than the creator, who is

[10] Robert Chambers, *Vestiges of the Natural History of Creation* (London: John Churchill, 1844).

[11] Richard Yeo, 'William Whewell, natural theology and the philosophy of science in mid nineteenth century Britain', *Annals of Science*, 1979, 36, 5: 493–516 at 497.

[12] Yeo, 'Whewell', 515. See also Laura. J. Snyder, 'It's *All* Necessarily So: William Whewell on Scientific Truth', *Studies in History and Philosophy of Science*, 25, 5, (1994): 785–807.

capable of understanding questions of design, intention and morality in the universe: a view which contrasts markedly with his earlier thoughts on the same topic expressed in his *Bridgewater Treatise*. Thus, in the revised Whewellian cosmos, man could aspire to approach a comprehension of the mind of God. Such an aspirational achievement would represent the pinnacle of human endeavour.[13]

Whewell's Response to *Vestiges of the Natural History of Creation*

At the other extreme, and despite his original tentative support for the ideas expressed in the nebular hypothesis, Whewell could not accept the deist position which would allow for the entire cosmos, including its sentient inhabitants, to unfurl as a consequence of adventitious evolutionary processes over which the creator exercised no immediate (and possibly not any) control and in which man is not accorded a special position in the process of creation.[14] Whewell was unhappy with the extensive use of such objectionable developmental ideas in *Vestiges*. He claimed to be particularly horrified by the lack of clear understanding of standards of scientific proof in the book:

> For the Vestiges has its hold on its readers, as it seems to me, by their want of the apprehension of the difference of the nature of truth and falsehood in science and philosophy; and this is a support which no argument can remove, but which on the contrary blunts the point of every good argument.

Whewell's dislike was shared by fellow Lancastrian Richard Owen (1804–1892) and his Cambridge colleague Adam Sedgwick (1785–1873): "You would be surprised to hear the contempt and abhorrence with which Owen and Sedgwick speak of the Vestiges."[15] From Owen, Whewell had also received the idea that the development of both extinct and extant vertebrate species depended on the existence of templates or archetypes thus "affirming that the ideal exemplar for the vertebrate animals existed

[13] Laura J. Snyder, 'Lord only of the ruffians and fiends'? William Whewell and the plurality of worlds debate', *Studies in History and Philosophy of Science*, 38 (2007): 584–592 at 587–588.

[14] See below and Chap. 6 for further discussion of Whewell's changing views on the nebular hypothesis.

[15] Whewell to Rev. F. Myers, 16 March 1845, in Stair Douglas, *Whewell*, 317–319.

as an idea in the Divine Mind".[16] This concept of a 'unity of plan' provided support for the concept of the nomological development of a diversity of species as opposed to the teleology of many natural theologians.

Arguably the primary cause of Whewell's dislike of the hypotheses unveiled in *Vestiges*, however, was the removal of the unique nature of man as God's special creation and the lack of any reference to final causes:

> I have attempted to show that, dim as the light is which science throws upon creation, it gives us reason to believe that the placing of man upon the earth (including his creation) was a supernatural event, an exception to the laws of nature. The Vestiges has, for one of its main doctrines, that this event was a natural event, the result of a law by which man grew out of monkey. As to the degree of resemblance between the author and the French physiological atheists, he uses reverent phrases: theirs would not be tolerated in England. But I do not recollect any reference to final causes in him.[17]

To Whewell, the avoidance of final causes was to be read alongside the slighting reference to "reverent phrases" as indications of doubt about the author's faith in the veracity of revealed religion. Clearly, the deployment of certain types of metaphysical argument provided a set of diagnostic data which led many readers, including Whewell, to believe that they could divine the heterodox leanings of the anonymous author.

Whewell's formal response to *Vestiges* took the form of a rehash of some selected excerpts of his own works in an 1845 book intended as a rebuttal of the elements of the *succès de scandale* with which he disagreed. This indirect attack, with the title *Indications of the Creator*, was published anonymously, presumably because Whewell did not wish to be seen as validating or giving equal authority to the allegedly pseudo-scientific content of *Vestiges* by attaching his own name to the counter-argument. One of the objectives of this work was to point out that some of the ideas used by the author of *Vestiges* were actually speculative conjectures rather than firmly established theories: the nebular hypothesis being a specific example. He also used extracts from his own treatises on scientific methodology to identify what he saw as deficiencies in the approach of the author of *Vestiges*.[18]

[16] Yeo, 'Whewell', 508.

[17] Whewell to Rev. F. Myers, 16 March 1845, in Stair Douglas, *Whewell*, 318.

[18] William Whewell, *Indications of the Creator* (London: John W. Parker, 1846).

Of the Plurality of Worlds: An Essay

In due course, Whewell wrote another anonymous essay in which he further revised his previously held positions on the probability of the existence of extraterrestrial organisms. Frederick Burnham has argued that Whewell's change of mind with respect to the 'truth' of pluralism as presented in *Of the Plurality of Worlds: An Essay* (1853) was motivated by his known preference for the concept of a creation which proceeded by the secondary operation of God's laws on the contents of the cosmos. In other words, his subsequent rebuttal of pluralism contains evidence of "his emphasis [...] on the nomological as opposed to the teleological form of the design argument".[19]

This partiality for nomological explanations actually aligned Whewell rather more with contemporaries like John Frederick William Herschel (1792–1871) than with the teleological approaches of a William Paley (1743–1805) or a Chalmers, despite his evident early admiration for the latter. It was a common characteristic of the emergent group of professional 'scientists' (a term actually coined by Whewell himself) and who represented a more open and questioning approach to explanations of natural phenomena.[20] And, of course, it is important in gaining an understanding of how Whewell's intellectual views were formed to note the more evident physical proximity and social intercourse that he experienced with Herschel, Sedgwick and his other peers who formed the overlapping communities of theologians and natural philosophers at Cambridge, whereas his contact with the father figures of natural theology was more remote in both time and space.[21]

In contrast to Burnham's explanation, John Hedley Brooke has argued that the main purpose of the *Essay* was to further refute the evolutionary hypotheses promoted by *Vestiges*:

[19] Michael J. Crowe, *The Extraterrestrial Life Debate 1750–1900: The Idea of a Plurality of Worlds from Kant to Lowell* (Mineola NY: Dover Publications, 1999), 266.

[20] Charles G. Gillispie (ed.), *Dictionary of Scientific Biography*, vol XIV (New York: Scribner, 1976), 292.

[21] Although William Paley died in 1805, Chalmers' career overlapped with that of Whewell, and they did indeed meet at, for example, the British Association meeting in Cambridge in 1833. Boyd Hilton, *The Age of Atonement: The Influence of Evangelicalism on Social and Economic Thought 1785–1865* (Oxford University Press, 2001), 60.

> His intentions do not stand out with great clarity—and yet that is precisely what one would expect if he was taking a left-handed swipe at the development hypothesis. Did Whewell calculate that to detonate the plurality of worlds was one way of exploding the presuppositions that had made *Vestiges* so dangerous? Did he, perhaps, give the game away in a letter to Forbes: 'I shall be surprised if anyone thinks that I give any support to the man of the *Vestiges*. It seemed to me that my book might have some value as a strong case exactly opposed to his?'[22]

This argument from the negative appears a little tentative at first sight and taken alone is not particularly convincing. Brooke could have provided further ammunition for his thesis if he had more clearly highlighted one specific quotation from *A Dialogue on the Plurality of Worlds*, appended to some editions of the *Essay*. Here Whewell made a more direct comparison with the main thrust of *Vestiges* whilst also coming very close to revealing the identity of the author of the *Essay*:

> If he were to try to connect me with an answer to that book, which went through two editions, under the title of Indications of the Creator, he would be nearer the mark. At least I adopt the sentiments of this latter book; and they agree with those of the Essay, as you may satisfy yourself by looking. In both works, the placing of man on the earth is regarded as an event out of the ordinary course of nature; directly the opposite of the doctrine of the Vestiges.[23]

There are therefore at least two available explanations for Whewell's change of mind on pluralism. The first posits a preference for nomological accounts of the process of creation which allowed him to place man at the pinnacle of sublunar creation with an aspiration to an understanding of the thoughts of the deity. The mind of man is thus an analogue for the mind of the creator in the same way as vertebrate body plans were seen to be templates for future development. An alternative argument, championed

[22] John H. Brooke, 'Natural theology and the plurality of worlds: Observations on the Brewster-Whewell debate', *Annals of Science*, 1977, 34, 3: 221–286 at 267. Brooke includes a quotation from a letter from Whewell to James David Forbes, 19 February 1854, also recorded in Isaac Todhunter, *William Whewell, D. D. Master of Trinity College, Cambridge. An account of His Writings with Selections from His Literary and Scientific Correspondence*, vol 2 (London: Macmillan, 1876), 400–401.

[23] William Whewell, *A Dialogue on the Plurality of Worlds* (London: John W. Parker and Son, 1853), 52–53.

by Brooke, is that Whewell sought in his anonymous work to remove the possibility of the existence of extraterrestrial life in order to defuse the evolutionary arguments of a controversial book which carried the seeds of a nascent atheism and which threatened man's special place in the creation. It is certainly clear that he objected strongly to both the style and content of *Vestiges*, and it also seems probable that the very popularity of the work was also a source of concern, as he recognised that its wide readership represented a genuine threat to his own metaphysical, religious and scientific beliefs.[24]

A third partial explanation, based on Whewell's idealism, has been offered by Yeo, although he conceded that his thoughts are "complementary to this [Brooke's] analysis":

> Whewell's rejection of the plurality of worlds is seen as an important aspect of his defence of an idealist epistemology. In asserting the unique place of man in the universe, he was able to rationalize the notion of a close affinity between the mind of man and the mind of God.[25]

On reflection, however, these explanatory accounts are contributory but at the same time insufficient as a totalising account of the evolution of his *Weltmodell* which had motivated Whewell to produce such a controversial reversal of his previous position that he again chose to publish it anonymously.[26] Clues to a more fundamental but also more subjective cause lie in an examination of recent affective changes which had taken place in Whewell's personal life and their relationship to the major arguments which he made in the *Essay*.

Taking the latter point first, it is clear from the structure and content of the 1853 book that Whewell, an ordained minister of the Church of England, wanted to preserve the fundamental Christian message of hope and salvation in the form of a special revelation communicated to "the race of man" alone:

[24] For a detailed discussion of the contemporary reception of *Vestiges*, see: James A. Secord, *Victorian Sensation: The Extraordinary Publication, Reception, and Secret Authorship of Vestiges of the Natural History of Creation* (University of Chicago Press, 2000).

[25] Yeo, 'Whewell', 506.

[26] A German word meaning 'world picture' which has previously been employed to describe both the cosmological models of physical scientists and the creation stories found in traditional religions. It thus encompasses the two key epistemologies under discussion here.

> Revealed Religion teaches us [...] that by his [God's] dealings with the race of man in the earlier times, and at various intervals, he made preparation for the mission of a special Messenger, whom, in the fulness of time, he sent upon the earth in the form of a man [...]. The arrival of this especial Message of Holiness, Judgment, and Redemption, forms the great event in the history of the earth, considered in a religious view, as the abode of God's servants.[27]

In making this argument Whewell was in part relying on his earlier assertion that the placing of man on earth was a special event and, in effect, the final cause of Creation. The status of man as the recipient of divine revelation was thus also explained.

At this point, Whewell had clearly distanced himself significantly from his earlier approval of the work of Chalmers, particularly on the question of the consolatory power of a cosmos teeming with life.[28] On the other hand, Whewell had also, to some extent, followed in his predecessor's footsteps in terms of his choice of theological reading matter. Both men were in part stimulated by a much earlier Christian apology: the *Pensées* of Blaise Pascal (1623–1662). Whewell mentioned Pascal's work in one of a series of letters to his friend and colleague Sir James Stephen (1789–1859): "I have some speculations of the nature of those of Pascal, though I have not referred to him in them."[29] The reason he gave for not including his unspecified "speculations" was a fear that "The public is ready to take offence at any speculations on religion which are at all out of the common track". We might also suspect that Whewell would have had some difficulty in promoting the irrational, rather depressing prose style and the Jansenist views of the French philosopher.[30] Pascal's *Pensées*, which focused on the need for faith rather than rationality in the process of Christian conversion, had been, however, evidently compatible with the *Weltmodell* of a convalescent Chalmers, and Whewell's interest in them is perhaps an indication of a similar turn to a fatalistic frame of mind combined with an increased need for religious belief at a time when he feared greatly for his wife's future.

[27] William Whewell, *Of the Plurality of Worlds: An Essay* (London: John W. Parker and Son, 1853), 44.

[28] See, for example, Whewell, *Dialogue*, 42–43.

[29] Whewell to James Stephen, 6 September 1853, in Todhunter, *Whewell*, vol 2: 379–380.

[30] Norman Wilde, 'Scepticism and Faith in the Philosophy of Pascal', *The Harvard Theological Review*, 9, 1 (Jan., 1916): 56–83 at 70–83.

Stephen, who had been appointed Regius Professor of Modern History at Cambridge, wrote a series of letters in which he persuaded Whewell to tone down some of his more pejorative anti-pluralist rhetoric in favour of a more measured and objective style of argument.[31] Towards the end of his reply to Stephen quoted above, Whewell characteristically switched away from his academic authorial voice to provide an update on family news, revealing just one of the "natural sorrows" which were subsuming his innermost thoughts at that time:

> I am greatly grateful for your kind sympathy in our sorrows. They are not single, for James Marshall [brother of Cordelia Whewell] has just lost a child. Well for us if we can find in our philosophy anything which can elevate and moralize our natural sorrows.[32]

Cordelia Whewell's Terminal Illness

The loss of his brother-in-law's child was by no means the only personal tragedy that had occurred or was in the process of occurring at the time when Whewell wrote the *Essay*. Like Chalmers, he suffered a series of family illnesses, including the loss of both his "dear brother-in-law Myers" and his wife's sister at a period in his life when he had also detected some signs of vulnerability in his personal health.[33] Whewell's cumulative experience of the loss of close relatives, although perhaps not abnormal for the period, occurred at a time when he was also anticipating the probable death, after a lengthy and debilitating illness, of his wife.

Cordelia Whewell (1803–1855, née Marshall) began to experience the symptoms of her ultimately fatal illness in 1844.[34] Writing to his wife at that time, Whewell expressed his concern that she speaks of "pain and weariness". He also talked about his admiration for her "'resignation'" which he felt he must now reluctantly adopt and "make common stock with". As noted in Chap. 1, an attitude of resignation, translated into heightened religiosity, was a common by-product of tubercular infection. Whilst the specific nature of her illness was never mentioned, it was known to produce some of the symptoms seen in Cordelia's slow decline:

[31] See Todhunter, *Whewell*, vol 2, 379–396 and Crowe, *Extraterrestrial*, 293–296.

[32] Whewell to James Stephen, 6 September 1853, in Todhunter, *Whewell*, vol 2, 379–380.

[33] Whewell to Dr Mackreth, 18 December 1851, in Stair Douglas, *Whewell*, 421. Whewell to Adam Sedgwick 13 April 1853 in Stair Douglas, *Whewell*, 424.

[34] Ibid., 307.

> I am grieved that you speak of pain and weariness, but they may lead to comfort and gratitude. For gratitude at any rate we have abundant cause; and if need be I must make common stock with you in your 'resignation,' which I used to exclaim against. I hope this change in my estimate of qualities is because I am become more reasonable.[35]

The language used here, including a projected response of "comfort" and "gratitude", reflects a prevalent form of theodicy in which God in due course will reward the virtues of a resigned approach to great suffering.[36] This trope, later to be combined with known patterns for coping with approaching death and bereavement, were understood by Whewell to be appropriate responses. At the same time a sense of guilt arose from his earlier impatience with Cordelia's resignation "which was by no means one of his favourite virtues", a reaction which he now felt to have been misplaced in view of the seriousness of her situation.[37]

Further evidence of his acceptance of the virtues of "blessed resignation and hope", which would prove to be a foreshadowing of the couple's approach to Cordelia's eventual death, can be found in Whewell's report of the death of Jane Marshall (his mother-in-law):

> After what you have heard already you will not be surprised by the news of Mrs. Marshall's death. She departed on Monday about 3 o'clock, having been long in a state of weakness and gradual sinking, which made the approaching end very plain. She was during her illness in a state of most blessed resignation and hope, and as she said 'unutterable peace' accompanied with constant prayer.[38]

Cordelia's symptoms evidently worsened towards the close of 1851. William's niece and biographer, Janet Mary Stair Douglas (c1830–1922), commented briefly on this period in his life: "The year 1851 closed in gloom. The letters of 1852 are few in number and show traces of anxiety and depression. Kreuznach was again recommended, and again resorted to."[39] William and his wife had visited Bad Kreuznach, a spa town in the

[35] Whewell to Cordelia Whewell, 18 June 1844, in Stair Douglas, *Whewell*, 309–310.

[36] Patricia Jalland, *Death in the Victorian Family* (Oxford University Press, 1996), 43.

[37] Whewell to Cordelia Whewell, 18 June 1844, in Stair Douglas, *Whewell*, 309–310.

[38] Whewell to Ann Whewell [his sister], 28 January 1847, in Stair Douglas, *Whewell*, 336–337.

[39] Whewell to Dr. Mackreth, 18 December 1851, in Stair Douglas, *Whewell*, 421.

Rheinland-Pfalz district of Germany, on several occasions to 'take the waters' and to seek some alleviation for the symptoms of Cordelia's illness. Kreuznach was also the location in which, during the summer of 1853, William wrote his *Essay*, at a time when the seriousness of his wife's long-term condition was a source of "extreme distress" as he experienced "the overwhelming ebbs and flows of hope and fear":

> I should probably have written to you if my mind had been more at ease. But with great anxieties about the persons dearest to us, pressing upon our minds, it becomes difficult to write any letters which are not absolutely necessary. I am full of the deepest trouble about my dear wife. I cannot look forward without a very heavy heart and extreme distress. I need not say to you that all this is for your private ear; for though these portions of the lot of our human nature have nothing in them which requires concealment, we cannot but wish the overwhelming ebbs and flows of hope and fear not to become matter of discussion to ordinary acquaintances.[40]

These sentiments were to be shared within a very limited contemporary circle and were not intended "to become matter of discussion to ordinary acquaintances". Nonetheless, they were simultaneously being preserved for futurity, representing a potential, but postponed, public display of apprehension and inner turmoil.

It is therefore clear that at the time of writing the *Essay*, Whewell was experiencing two powerful but contrasting affects. On the one hand, he was increasingly anxious about his wife's declining health, but on the other he had absorbed conventional contemporary expectations of what a good death should look like and that it should, if properly performed, lead to a celestial reward.

Stair Douglas found ample evidence of an ongoing consolatory narrative in the form of a report, which a later reader might take as having been idealised for external consumption, of a deathbed conversation between William and Cordelia in which they appear to be attempting to adopt an almost Christ-like resignation, echoing the passage from Matthew's Gospel which related the events in the Garden of Gethsemane prior to the arrest of Jesus:

> In praying by her side, I said, more for myself than for her, 'If this cup may not pass from me except I drink it' ... and I could not finish the sentence,

[40] Whewell to Dr. Mackreth, 6 October 1851, in Stair Douglas, *Whewell*, 425–426.

> but she finished it and said, 'Thy will be done.' It is very hard to say, my dear sister, but I must learn the lesson of you as soon as I may.
>
> She said, 'Do not put upon my grave-stone anything that is merely sad and sorrowful.' I replied, 'We are not as those who sorrow without hope.' 'Yes,' she said, 'not without hope;' and I added, 'Your own device which you gave to Susan: "We live to die, we die to live."[41]

The apparent calm acceptance of Cordelia's forthcoming demise echoes several prototypes for this kind of deathbed Christian confession of which the author must have been aware. One year before Whewell published his *Essay*, for example: "Dorothea Palmer, sister of the first Earl of Selborne, died of consumption [probably tuberculosis] in 1852 at the age of about 23." During her fatal illness, she took comfort from "John Keble's consolation poetry, *The Christian Year*", which provided her with a pattern to be followed:

> Dorothea Palmer's death was a story of absolute resignation to God's will and a strong but simple faith. [...] Her brother Revd George Horsley Palmer hoped to follow Dorothea's example when his turn came to die: 'I cannot wish or desire more blessedness—her faith—her meekness—her brightness and resignation indeed deprives death of every sting.'[42]

The *Essay*: Whewell's Response to His Wife's Illness and a Rebuttal of Chalmers

Chalmers' response to an existential crisis had been to adopt and promote Evangelism and pluralism, both concepts which were prominent in the cultural context of the early nineteenth century. His reply to the Psalmist's enquiry as to the value placed by God on an individual soul was to insist on the omnipotence of the deity and to opine that every individual being, human or otherwise, would be the recipient of theistic care. Whewell, however, was writing his *Essay* at a time when religion was moving away from the potential deism of natural theology and towards an environment which combined slowly increasing scepticism with an expanding palette of

[41] Whewell to Mrs. F. Myers, 23 December 1855, in Stair Douglas, *Whewell*, 448–450. The quotation which alludes to events in Gethsemane was taken from Matthew 26:42. "We live to die, we die to live" is a paraphrase of Romans 14:8.

[42] George Horsley Palmer to William Palmer, 15 July 1851, in *Selborne Papers* (Bodleian Library, Oxford) MS Eng. misc. c. 691, folios 3–4, quoted in Jalland, *Death*, 43.

personal pieties. He now disavowed his earlier support for "the great Preacher of Scotland":

> But I may say, on the other hand, that I have found religious persons, who, returning to the perusal of Chalmers's Astronomical Discourses, now that the first burst of their popularity arising from their great eloquence, is past, have found his suppositions and pictures painfully bold and his views very far from consolatory.[43]

Whewell preferred to console himself and his reader by radically limiting the possible existence of extraterrestrial life in order to preserve the unique status of "Human Society":

> God thus directly instituted Human Society; gave man his privileges and his prospects in such society; placed him far above the previously existing creation; and, endowed him with the means of an elevation of nature entirely unlike anything which had previously appeared.[44]

The case which has been made in the preceding paragraphs is that in Whewell's case the concept of pluralism was not only informed by idealist metaphysics but also by a very personal concern for his wife's eternal well-being and thus for a version of the universe in which Christian soteriology could be accommodated. It had, in his *Weltmodell*, become a prerequisite of the preservation of the soul that the crucifixion and resurrection were events unique to earth as Niamh Brown has pointed out:

> The central Christian proof of God's love for humanity is Jesus' death on the cross to save mankind. This redemption, Whewell suggests, 'shows, of course, that God has an especial care of man' which gives Earth the status of the 'selected [...] theatre of redemption', and places it above 'being on a level with any other domiciles'.[45]

To support this existentially charged argument, Whewell used the first four chapters of his response to Chalmers' earlier thesis to lay out what were in effect his religious, moral and, in particular, personal issues with

[43] Whewell. *Dialogue*, 42–43.

[44] Whewell, *Essay*, 277.

[45] Niamh Brown, 'Devotional Cosmology: Poetry, Thermodynamics and Popular Astronomy, 1839–1889', PhD thesis submitted to the University of Glasgow (2016): 138.

the possible existence of extraterrestrial life before he moved on in the next six to state several more evidently scientific, for which read ostensibly objective, arguments. The *Essay* is therefore almost two books in one. Although a form of Christian apologetic was amongst the principal purposes of the *Essay*, Whewell had also in fact produced a comprehensive review of the empirical evidence for and against the concept of pluralism. As we saw in Chap. 2, most previous writers had approached the topic with a thin leavening of astronomy underlying imaginative attempts to bolster or oppose established religion. All in all, it can be seen in retrospect that, given the lack of definitive data, Paley's decision not to employ astronomy as a potential proof of God's existence was indeed well judged.

Astronomy, which Whewell acknowledged "claims a sort of dignity over other sciences, from her antiquity, her certainty, and the vastness of her discoveries", had in his view shown that the universe is characterised by both vast space and deep time:

> And as the infinite extent which we necessarily ascribe to space, allows us to find room, without any mental difficulty, for the vast distances which astronomy reveals, and even leaves us rather embarrassed with the infinite extent which lies beyond our farthest explorations; so the infinite duration which we, in like manner, necessarily ascribe to past time, makes it easy for us, so far as our powers of intellect are concerned, to go millions of millions of years backwards.[46]

Having raised the possibilities of "infinite extent", Whewell then decided *au contraire* to restrict the opportunities for pluralist biology by attempting to minimise the physical space available for such developments. He argued "that the Sun is the largest Sun" in the universe and that the earth could be the largest solid planet, with the intention of severely restricting the extent of interstellar terra firma:

> So far as we yet know, the Sun is the largest Sun among the stars; and we shall attempt to show, that the Earth is the largest solid opaque globe in the solar system. Some System must be the largest and most finished of all; why not ours? Some planet must be the largest planet; why not the Earth?[47]

[46] Whewell, *Essay*, 101, 68.
[47] Ibid., 160.

Similarly, in a chapter on 'The Nebulae', he claimed, on the basis of John Hershel's observations of the Magellanic Clouds, that all 'unresolvable nebulae' are indeed nebulous and do not constitute thousands or even millions of remote star systems which might be locations for life, thus also refuting the relatively recent claims of nebular resolvability made by Lord Rosse.[48]

One of Whewell's key points appeared in Chapter VI: 'The Argument from Geology'. This geological argument was actually a further discussion about the duration of past time. Starting with a review which drew from his personal knowledge of geology, he concluded that it was difficult to avoid this conclusion: "There is therefore nothing, in what we know of the Cosmical conditions of our globe, to contradict the Terrestrial evidence for its vast antiquity, as the seat of organic life." The comparatively recent advent of humans, however, provided him with a counter argument to the astronomical objection:

> When, therefore, Geology tells us that the earth, which has been the seat of human life for a few thousand years only, has been the seat of animal life for myriads, it may be, millions of years, she has a right to offer this, as an answer to any difficulty which Astronomy, or the readers of astronomical books, may suggest.[49]

His proposition was that the presence of human beings occupies such a tiny proportion of the total history of the Earth that it probably has not happened and indeed may never happen on other planets, which allowed him to assert the special nature of humankind and to argue that there had been only a single act of divine revelation and redemption:

> He has bestowed upon this small part of His creation, mankind, this special care;—He has made their period, though only a moment in the ages of animal life, the only period of intelligence, morality, religion.[50]

The main thrust of this argument against pluralism, based on the epistemologically unusual combination of limited real estate and a lengthy timeline of which the existence of man only represented a small fraction,

[48] Ibid., 113–139. See Chap. 7 for an extended discussion of the observations made with Rosse's telescopes.

[49] Ibid., 102.

[50] Whewell, *Essay*, 103.

was to be subsequently taken up by several commentators.[51] For Whewell, this argument constituted a kind of special pleading which, in a marked contrast with Chalmers' omnipresent pluralist God, allowed him to preserve the concept of a personal theistic deity who can more than adequately care for the physical, moral and spiritual welfare of a limited population of human souls.

Reactions to Whewell's *Essay* were, to say the least, somewhat mixed despite its overt apologetic intent. John Herschel, clearly already aware of the identity of the author, doubted whether Whewell was sincere in his conclusions, whilst incidentally providing some independent confirmation of the negative state of his *Weltmodell*:

> The essay on the 'Plurality of Worlds' (attributed to him, though published anonymously), can hardly be regarded as expressing his deliberate opinion, and should rather be considered in the light of a *jeu d'esprit*, or possibly, as has been suggested, as a lighter composition, on the principle of '*audi alteram partem*' undertaken to divert his thoughts in a time of deep distress.[52]

In general, the *Essay* did not seem to chime well with much previous thinking on the topic of extraterrestrial life, and it attracted the ire of many supporters of pluralism including the Scottish astronomer and physicist David Brewster (1781–1868).[53] Brewster began a lengthy campaign against Whewell's publication, seeking to defend a vision of a universe in which God had created a plenitude of life-bearing planets as compared with the somewhat dismal vision of a single inhabited world "unoccupied for millions of years".[54] One of the intriguing elements of this debate for more recent historiography is the unexpected religious alignments of the protagonists. Brewster, a theologically conservative Scottish Evangelical, argued in favour of a plethora of evolutions which, though aligned with some tenets of natural theology, had also been advanced as an argument against Christianity. On the other hand, Whewell, a liberal Anglican,

[51] Crowe, *Extraterrestrial*, 457–475.

[52] John F. W. Herschel in *Proceedings of the Royal Society*, vol XVI, LX–LXI. Quoted in Todhunter, *Whewell*, vol 1, 187. '*Audi alteram partem*' can be translated as 'the other side should be heard'.

[53] This debate is covered at some length in both Crowe, *Extraterrestrial*, 300–304, and Brooke, 'Brewster-Whewell'.

[54] Ibid., 278.

wanted to bolster the more traditional account given by revealed religion which posited the planet earth as a 'special case'.

The attitude of Brewster provides an interesting parallel with the earlier case of Thomas Chalmers, who was Brewster's contemporary and occasional colleague. Chalmers, who had been born into a similar geographical and religious milieu, and who shared his proselytising tendencies, had likewise promulgated the concept of a populous universe as evidence of God's creative and redemptive powers following his 'Pauline' conversion. As Brooke commented in his dissection of the debate: "For historians of science it provides an interesting case-study in the 'openness' of astronomical data to interpretations strongly informed by metaphysical and theological commitment."[55]

Whewell and Politics

By way of contrast with the changing theological stance which we have just noted, it is difficult to discern any significant shifts in Whewell's political views in the same time period. Early in his career, he had displayed a certain level of conservatism in the matter of freedom of worship in Trinity College, but on the other hand he had also supported the Catholic Emancipation Act and the repeal of the Test and Corporations Act of 1828, so we are led to the conclusion that his views were neither of a particularly Tory nor of Radical persuasion. Instead, he generally pursued a middle path of being a "constitutional conservative" and an "old Whig" throughout his life:

> As a young man he seemed inclined to take a Liberal line, for he opposed a petition from the University against the Roman Catholic claims in 1821 [...]. But in those stormy days of the Reform Bill, when so many ancient friendships were destroyed he took no decided line: and latterly he abstained from politics altogether.[56]

[55] Brooke, 'Brewster-Whewell', 279.

[56] Whewell engaged in a vigorous debate with the Reverend Connop Thirlwall about the requirement for compulsory College Chapel attendance for Dissenters. But he did not hold any animosity towards Thirlwall and was pleased that the "Whig Ministry" had awarded him a "valuable living in Yorkshire". See Whewell to C. Thirlwall, 28 May 1834, and Whewell to A. Whewell, 19 December 1834, in Stair Douglas, *Whewell*, 162–168. Laura J. Snyder, *Reforming Philosophy: A Victorian Debate on Science and Society* (Chicago: University of Chicago Press, 2006), 216, 223. John W. Clark, *Old Friends at Cambridge and Elsewhere* (London: Macmillan, 1900), 10.

Although his fellow Cambridge graduate, John Willis Clark (1833–1910), may have viewed Whewell's political views as studiedly neutral, it is, in fact, clear that he would not have become the master of Trinity College, and the husband of Cordelia, if he had not been seen as sufficiently 'sound' by his peers and his predecessor "[Christopher] Wordsworth, an unreconstructed Tory" who refused to resign "until the election of a Conservative government in the autumn of 1841", thus paving the way for Whewell's succession:

> Whewell was recommended to the Queen by the prime minister, Peel, and duly appointed master. Simultaneously, Whewell had advanced his own career and obtained a position for which celibacy was not a prerequisite.[57]

The change of Whewell's *Weltmodell*, which we have seen taking place in and around 1853, did not, if we take Clark's comments as a guide, lead to any noticeable changes in his political views: an arena which carried less-affective freight than the existential concerns which appear to have informed much of his theology.

In this context, it is also worth noting that Simon Schaffer has offered an analysis of Whewell's rationale for writing the *Essay* in which the claimed uniqueness of man should be further extended to identify the centrality of Britain as a balanced and conservative polity, avoiding the excesses of "extreme German idealism" on the one hand and the "'extravagantly democratic' regimes in France" on the other. It therefore might not be a reading too far to say that Whewell wanted to position England and, even more specifically, Cambridge at the core of his imperium. "Astronomical and geological elitism went with political chauvinism and its defence".[58]

[57] Nicholas W. Gillham, *A Life of Sir Francis Galton: From African Exploration to the Birth of Eugenics* (Oxford: Clarendon Press, 2001), 41.

[58] Simon Schaffer, 'The History and Geography of the Intellectual World: Whewell's Politics of Language', in Menachem Fisch and Simon Schaffer (eds.), *William Whewell: A Composite Portrait* (Oxford: Clarendon Press, 1991), 201–232, at 205–207.

Cosmology as Consolation

Whewell had clearly made some significant transitions during his lifetime in the fields of religion, metaphysics and natural philosophy. As a supporter of natural theology in the 1830s, he initially followed the route laid out earlier by Chalmers. In the late 1840s, he played down the possibility of intelligent beings existing elsewhere in the universe to support his emerging thesis in which human beings were to be seen as both the peak achievement and the purpose of the creation. By 1853, following the loss of close relatives and in parallel with the serious, ultimately fatal, illness of his wife, he decided to publish a controversial text, seen by others as not congruent with his preceding views or prevailing opinion, which argued for a totally anti-pluralist position. By denying the existence of extraterrestrial life altogether, he was able to preserve or even enhance the concept of personal salvation for Cordelia and himself.

In his widely circulated *Essay*, Whewell chose to restore to his version of the universe a single and personal revelation. Whewell's changed views were supported by some strong empirical arguments: once he had decided to reverse his direction of travel, he was able to show that there are many pointers to the improbability or, at the very least, relative scarcity of other populated planets. The overall tone of the *Essay*, combined with it having been published anonymously, suggested that its main purpose was as an exercise in which Whewell had privately decided to reverse some well-known elements of his previous cosmological thinking which, it now seemed to him, militated against a full acceptance of orthodox soteriology at a time of personal crisis: "a consummation Devoutly to be wish'd".[59]

The publication of the *Essay* also coincided with a noticeable shift in Whewell's views on metaphysics. This change was subtler but nonetheless consistent with his evolving *Weltmodell*. It entailed a further maturation of his conception of human thoughts as "shadows" of "Divine Ideas". This was a logical extension of his earlier thinking about the way in which the ability to comprehend God's thoughts is proof of man's unique position in creation as Snyder has noted:

> As we have seen, by the mid-1850s Whewell came to found his epistemology on a theological base. Since our Ideas are "shadows" of the Divine Ideas, to see a law as a necessary consequence of our Ideas is to see it as a

[59] William Shakespeare, 'Hamlet, Prince of Denmark', Act 3, Scene 1.

> consequence of the Divine Ideas exemplified in the world. Hence the more we idealize the facts, the more difficult it will be to deny God's existence.[60]

Bearing in mind that Whewell had been a very able scientist, administrator and author, it is not hard to imagine that he was well aware that some of his correspondence would ultimately appear in the public domain, and he therefore may well have carefully constructed his literary persona to be congruent with his public utterances as a minister and master of Trinity College. Nonetheless, it seems that underlying his public profile there was indeed a strong bond of affection between Whewell and his first wife. It is not difficult to perceive that the long periods of her illness spent in Kreuznach and Lowestoft, during which he was often physically apart from his wife, had placed a heavy strain on Whewell which Stair Douglas has documented. As noted above, John Herschel had opined that these and other affective events provided a sufficient cause for Whewell's otherwise unexpected reversal on the topic of pluralism.[61]

Affective life events can be seen to have resulted in significant changes in the religious and metaphysical views of the people whom we are discussing in this book which in turn influenced their motivations for, and interpretation of, their astronomical projects. In most cases, their reactions resulted in increased religiosity. A common nineteenth-century trope, which is still present as a coping mechanism today, was to take a more optimistic view of the probability of the soul's survival beyond death thus enabling the bereaved to rendezvous with their lost partners or relatives in a presumed future life. As an instance of this cognitive response to the affect of loss and bereavement, we can perhaps take an excerpt from a message sent to Herschel by Whewell in 1861 on the occasion of a similar occasion for grief and reflection: the death of one of his nieces who had married into Herschel's family. In this letter, he endeavoured to console the 68-year-old polymath by reminding him of the material and social rewards, the "sources of earthly happiness", which are available in the here and now together with a forward-looking prospect in which "the separation cannot be for long". Whewell made the case that, as they were both ageing men, they will shortly be reunited with a majority of the people

[60] Snyder, *Reforming*, 92.
[61] Stair Douglas, *Whewell*, 421–428, 435, 442.

whom they had known and loved during the few short years spent in this sublunary sphere.[62]

Religious consolation was not a universal reaction to negative affect, however, and we shall, in the next chapter, examine one notable exception in the form of the astronomer Richard Anthony Proctor (1837–1888), who reacted to theological challenges by evincing increasing religious doubt. A potentially important difference which marks out his case history from the general trend noted above is that his familiarity with, and even acceptance of, the more materialist thought processes of a number of his contemporaries. It is probably also relevant to note that the triggering event leading to his change of views, was not in any way life-threatening.

[62] Whewell to John Herschel, 27 January 1861, in Stair Douglas, *Whewell*, 520–521.

CHAPTER 5

Richard Proctor and Private Judgement

Whilst round the chariot's way
Innumerable systems rolled,
And Countless spheres diffused
An ever-varying glory.

Percy Bysshe Shelley, *Queen Mab*[1]

Throughout the nineteenth century, there were many commentators who wrote or spoke on the topic of pluralism and the likelihood of extraterrestrial life. For the most part, they followed the spirit of the major intervention made by Thomas Chalmers (1780–1847) and were in favour of the concept. There were, however, a few exceptions, that is, opponents of pluralism, constituting an interesting minority which in due course included William Whewell (1794–1866). The latter group was joined later in the century by Richard Anthony Proctor (1837–1888), a well-known populariser of astronomy, who, following an intense internal debate, reversed a theological position which had been prompted originally by an

[1] Percy B. Shelley, *Queen Mab: A Philosophical Poem* (London, 1813), 14. The first edition was printed by Shelley for private circulation, and the unused copies were deposited at William Clark's bookshop.

H. Carlton, *Cosmology and the Scientific Self in the Nineteenth Century*, https://doi.org/10.1007/978-3-031-05280-4_5

affective challenge to his *Weltmodell*.[2] At the time of this unconversion, he also rethought his views on the relative epistemic authorities of religion and science. This interior review, which took place in response to a specific event and its reception by his peers and in the public press, also caused him to re-evaluate his position on the probability of extraterrestrial life, although not as radically as Whewell, and following an entirely different, or indeed opposing, narrative arc.

Proctor's early life and career are not well documented. Some of the best available sources of information on his life and thoughts are the obituaries written by his colleagues and his own published output in the form of books, articles and his many contributions to a magazine called *Knowledge*, which he founded in 1881 and edited until his death.[3] The relevant archive resources also include a series of vitriolic epistles to the Astronomer Royal and various journals in which, having been rejected as a candidate for the Royal Astronomical Society's Gold Medal, he proceeded to vent his spleen with fellow members of the Society.

Proctor also engaged in several disputes about funding proposals and astronomical techniques culminating in "A vigorous war of words [which] was carried on in *The Times*". This particular disagreement concerned the proposed observations of the 1874 transit of Venus, and it led to a note expressing "strong disapprobation" of Proctor's written comments being published by the Council of the Society and the termination of his brief occupancy of the office of Secretary.[4] These incidents provide pointers to one of the most persistent features of Proctor's persona: his outspokenness when he vigorously opposed any idea or initiative with which he was in disagreement combined with a noticeable resentment of authority. We should, however, be a little cautious in summarising his personality solely on the basis of published correspondence. In a passage which seems to reflect some genuine warmth, even though it might alternatively be read as merely conventional gallantry, his friend and colleague Edward Clodd (1840–1930) recollected: "To the outer world a polemic, ever delighting

[2] A German word meaning 'world picture' which has previously been employed to describe both the cosmological models of physical scientists and the creation stories found in traditional religions. It thus encompasses the two key epistemologies under discussion here.

[3] Charles G. Gillispie (ed.), *Dictionary of Scientific Biography*, vol XI (New York: Scribner, 1975), 162–163.

[4] John L. E. Dreyer and Herbert H. Turner, (eds.), *History of the Royal Astronomical Society: Volume 1 1820–1920* (1923 repr. Oxford: Blackwell, 1987), 176–184, 251. Proctor acted as secretary from 1872 to 1874, a brief period compared with that of both his predecessor and his successor (see table on page 251).

in controversy and fulminating against abuses in high places, a more tender-hearted mortal never lived."[5]

We do know that his mother, who had expressed a wish for him to pursue a theological vocation, died while he was at the University of Cambridge. The loss of his mother, "to whom he was devotedly attached", combined with the commencement of an interfaith relationship with Mary Mills, "a young Irish lady" whom he married shortly thereafter, preceded a noticeable dip in his academic attainment. He graduated as the 23rd wrangler in 1860: "a degree which greatly disappointed his friends, many of whom had already recognised his remarkable talent".[6] A domestic financial crisis caused by the failure in 1866 of a New Zealand "bank in which he was the second largest shareholder [...] leaving him with a staggering debt of 13,000 pounds" led to his decision to take up self-employment as a writer of popular books on astronomy and other scientific themes in order to support his family.[7] His limited and unpredictable income may well have been a further stimulant for his heated disputes with more fiscally secure correspondents who occupied positions of relative power.

Proctor's Conversion to Catholicism

It eventually became evident that at some time after his marriage he had converted to Catholicism. Clodd, himself a Unitarian '"sitting under" Dr James Martineau, believed that it had been the effect of "the death of a darling boy" in 1863 which had precipitated Proctor's "perversion" and that he was "seeking consolation in the Roman Catholic Communion":

> His perversion came upon me as a surprise, and there resulted a correspondence between us in which I got the worst of it. Let me explain. At that time I was reposing on "the soft feather-bed for falling Christians," as Unitarianism has been caustically defined. [...] Proctor had his feet on the rock of Saint Peter; mine were on the shifting sands of Theism.[8]

[5] Edward Clodd, *Memories* (London: Chapman and Hall, 1916), 58.

[6] Arthur C. Ranyard, 'Richard Anthony Proctor', *Journal of the Royal Astronomical Society*, XLIX, 4 (Feb. 1889): 164–167.

[7] Michael J. Crowe, *The Extraterrestrial Life Debate 1750–1900: The Idea of a Plurality of Worlds from Kant to Lowell* (Mineola NY: Dover Publications, 1999), 369. Bernard V. Lightman, *Victorian Popularizers of Science: Designing Nature for New Audiences* (University of Chicago Press, 2007), 301–302. Dreyer and Turner, *History*, 167–168.

[8] Clodd, *Memories*, 58–60. The idea that Unitarianism is a "feather bed for falling Christians" was also used by Charles Darwin, who attributed it in turn to his grandfather Erasmus.

Proctor's religious views were discussed at some length in a series of letters which passed between the two colleagues in July and August 1870. Clodd's presumption of a causal relationship between the death of Proctor's son, which led in turn to a lengthy period of ill-health, and his subsequent conversion should be read alongside the other contextual and cultural factors which impinged on him at about this time, including the shifting theological environment in which he was living and the financial disaster of 1866.[9]

We should also note Proctor's own comment in the last letter of this series that it was "nearly 3 years since I became a Cath". Thus, there was, according to Proctor, a significant interval between the tragedy of losing the boy and his conversion to the Catholic faith which indicates that, as in other cases discussed in this monograph, there can be a delay of several years between the locus of a *Weltmodell* dislocation and an externally detectable theological shift. Although the seeds of change may have been sown at a particular point in time, they sometimes required a lengthy gestation period. In the same letter, Proctor revealed that it had taken him a long time to completely come to terms with Catholic doctrine: "It seems to me now as though 3 years ago I had comparatively slight reason for accepting the churches authority, & that the reasons I have now are overwhelmingly strong."[10]

The discussion of Proctor's conversion had begun with a letter in which Clodd, whilst thanking him for a copy of his recent book, entitled *Other Worlds than Ours*, pointed out almost in passing that he had gained the impression from the final chapter that Proctor might be "willing to abrogate private judgement when I should feel bound to exercise it". This comment led to Proctor's subsequent admission of his acceptance of Catholicism and hence his rejection of private judgement and acceptance of Papal infallibility: two key doctrines of the Catholic Church. But the description of this process, especially in relation to private judgement, gives the reader considerable grounds to suppose that the effect of bereavement had faded somewhat and that by 1870 Proctor's faith was potentially vulnerable to ongoing doubts which he was trying hard to ignore:

[9] Lightman, *Victorian popularizers*, 301.

[10] Proctor to Clodd, 27 July 1870, Leeds University, Brotherton Library Special Collections, *Clodd Correspondence*.

> Whether I was in those times more changeful than most men I do not know; but certain I am that there was a continual fluctuation in my views, for I was at the pains to record from time to time the condition of my mind as respects religion. That such fluctuation should eventually lead to the rejection of Christian doctrine altogether seems to me a perfectly logical termination to the mental conflict. [...] I look at the so-called freedom of private judgement as a delusion and a snare![11]

Proctor's religious views did indeed become increasingly sceptical over the course of time, as noted by Clodd in his memoirs: "However, the illusions begotten of his unbalanced emotions were dispelled in later years, and he died an Agnostic."[12] It is evident from the duration and uncertainty of the preceding and ongoing interior debate that such an intellectual weighing of the arguments had led to a decision which could very easily be reversed by further mental struggle, as was in due course to be the case. Thus, Proctor eventually succumbed to the "so-called freedom of private judgement" despite his earlier self-denial.

Proctor and Pluralism

Turning to the specific topic of Proctor's attitude towards pluralism and the evidence of intersection with theological motifs, it is clear that he was initially an enthusiast for a universe widely populated with alien lifeforms. *Other Worlds* (subtitled 'The Plurality of Worlds studied under the light of recent scientific researches') was his "first major success" as an author.[13] This book, which had prompted the flurry of correspondence between him and Clodd, was influenced by, and was in many ways a response to, the preceding debate between the astronomer David Brewster (1781–1868) and Whewell on the topic of pluralism. It came down fairly solidly on the side of the former, whose views were aligned with those of Brewster's sometime colleague and fellow Scot Chalmers. Proctor extolled the concept of "the immense varieties of the forms of life" to be found in the universe:

[11] Proctor to Clodd, 27 July 1870, Clodd to Proctor, 11 July 1870, Proctor to Clodd, 12 July 1870, *Clodd Correspondence*.

[12] Clodd, *Memories*, 60.

[13] Crowe, *Extraterrestrial*, 369–370.

> But it is when we consider recent discoveries in their relation to the existence of other worlds, when we attempt to form a conception of the immense varieties of the forms of life corresponding to the innumerable varieties of cosmical structure disclosed by modern researches, that we recognize the full significance of those discoveries.[14]

Proctor was also clearly cognisant of the major theological concern which had informed some earlier writings on this topic: the question of how a single deity could provide "care and solicitude" for individual human beings at the same time as bestowing "infinite love" on "uncounted millions". He echoed some doubts which had been expressed on the various attempts by his predecessors to justify their personal positions on this issue. "I doubt almost, when I judge from the comments which have been made on this part of the works of Chalmers, Whewell, Brewster, and others, whether a single reader of those works has found the religious views of any one of their authors congenial with his own."[15]

Having highlighted this trap for the unwary writer, Proctor then somewhat perversely fell headlong into it, producing a lengthy and highly speculative essay in which he attempted to provide a scientific explanation of the way in which a truly supernatural being, possessing extremely powerful vision and the capability of travelling much faster than the speed of light, could appear to be omnipresent in both time and space. He based this hypothesis quite closely on a brief but entertaining monograph entitled *The Stars and The Earth* (1846), which Proctor erroneously believed had been produced anonymously but was in fact clearly the work of a German amateur astronomer, Felix Eberty (1812–1884).[16] Proctor was aware of several objections to Eberty's original thesis which he now attempted to resolve whilst simultaneously preserving the operation of known physical laws:

[14] Richard A. Proctor, *Other Worlds than Ours* (1870 repr. New York: D. Appleton, 1886), 21.

[15] Proctor, *Other Worlds*, 306.

[16] One of the more intriguing attempts to reconcile astronomical scales with an omnipotent deity, which to some extent foreshadowed theoretical concepts that were expanded in due course, can be found in: Felix Eberty, *The Stars and the Earth* (London: H. Bailliere, 1846). It was studied with interest by, amongst others, William Whewell and Albert Einstein. Proctor, *Other Worlds*, 312, is unaware of the authorship of the monograph.

> To sum up, we find ourselves led to the belief that, while intervention with the operation of natural laws is unnecessary, all the worlds existing throughout space are, in a very definite and special manner, watched over and controlled by an omnipresent, omnipotent, and omniscient Being; that before Him the infinite past and the infinite future of the universe are at all times sensibly present.

The end result is a very mechanistic view of the operations of the "God of Nature" which conveniently managed to avoid any discussion of how his "omnipresent, omnipotent, and omniscient Being" is able to understand or influence the inner workings of the human soul.[17]

By 1875, however, Proctor's views on the probability and universality of extraterrestrial life had significantly altered and he deployed a much more "Whewellite" thesis as noted by Michael Crowe:

> Readers of the first three essays in Proctor's *Our Place among Infinities* and of the first essay in his *Science Byways* could scarcely doubt that his views had undergone a major transformation since 1870.[18]

In these two new books, Proctor concurred with Whewell by expressing several objections to the universality of life on other planets. On the one hand, he had become increasingly convinced that Jupiter and Saturn were still in the throes of condensing from a primeval nebula and were consequently both too warm and too diffuse to support any form of life which we would recognise.[19] This idea had first been outlined in *Other Worlds* and had received limited approbation from John Frederick William Herschel (1792–1871) at that time.[20] Proctor also recapitulated the key chronological argument made by Whewell in *Of the Plurality of Worlds: An Essay* in which his predecessor claimed that during long eras of past and future time there will be no life or only very limited life forms on any given world:

> Whether we turn our thoughts to planet, sun or galaxy, the law of nature (recognised as universal within the domain yet examined), that the duration

[17] Ibid., 319–321, 333–334.

[18] Crowe, *Extraterrestrial*, 373.

[19] Richard A. Proctor, *Our Place among Infinities* (London: Henry S. King, 1875), 11, 66–67.

[20] Herschel to Proctor, 11 May 1870, Royal Society Herschel MSS HS 14.123.

> of life in the individual is indefinitely short compared with the duration of the type to which the individual belongs, assures us, or at least renders it highly probable, that in any member of any of these orders taken at random, *it is more probable that life is wanting than that life exists at this present time.*[21]

Although Proctor gave the reasons for his change of mind as being his evolving views on the constitutions of the gas giants and a reconsideration of the viability of Whewell's chronological argument, it is apparent that there was another key factor influencing his *Weltmodell*: he had effectively decided to divorce science from religion and treat them as two distinct and incommensurable epistemologies which had the no doubt welcome side effect of liberating him from the tortuous mental gymnastics which he had subjected himself to at the conclusion of *Other Worlds*. It is also clear that even in 1870, before the formal revocation of his Catholic faith, his conviction that both space and time are infinite meant that he could not accept the conventional cosmogony of revealed religion:

> I have spoken of the creation of all things, whereas, in the sense in which men can alone interpret such words, we cannot reasonably conceive that there ever was a creation; and I have spoken of the beginning, whereas we cannot conceive that there ever was a beginning in the sense implied.[22]

Another way to consider this disconnect is that Proctor now found it impossible to continue to bridge the gulf between the language and actions of the two cultural contexts to which he had hitherto subscribed: those of the Catholic Church and the semi-professional world of astronomy. He had previously been able to maintain a foot in both camps more or less successfully, but it had become too big a gap for him to be able to continue to negotiate. Proctor's 'divorce' also suggests the after-effects of a significant shift in public perceptions of the relative epistemic authorities of science and religion since Whewell had published his *Essay* in 1853.[23]

The specific event, or sequence of events, which drove Proctor to his theological schism has been briefly mentioned in extant historiography.

[21] Richard. A. Proctor, *Science Byways* (London: Smith, Elder, 1875), 34. Italics in original.

[22] Proctor, *Other Worlds*, 308.

[23] Frank M. Turner, 'The Victorian Conflict between Science and Religion: A Professional Dimension', *Isis*, 69, 3 (Sep., 1978): 356–376 at 370, 374. Two of the significant events which had occurred in the meantime were the publications of Darwin's theory of evolution in 1859 and a theologically controversial edition of *Essays & Reviews* in 1860.

Bernard Lightman, for example, has commented: "According to one obituary account, Proctor severed his connection to Catholicism in 1875 when he was told that some of his astronomical theories were not in conformity with the teachings of the Catholic Church."[24] More specifically, it is clear, based on an examination of his personal correspondence and contemporary newspaper articles, that Proctor's unconversion was a response to events surrounding an address given by the Irish protestant physicist John Tyndall (1820–1893) and some of the critical responses which it had received.

John Tyndall and the *Belfast Address*

Tyndall, in his role as newly installed President, delivered the address which prompted Proctor's apostasy to the British Association for the Advancement of Science in Belfast in August 1874. Some commentators were evidently quite offended by the content of the speech which was indeed generally viewed by his audience as being rather more controversial than was normally expected at an inauguration. "Probably no single incident in the conflict of religion and science raised so much furor"[25] Despite its apparently provocative content, however, the range of reactions to Tyndall's presentation was somewhat more heterogeneous and reflexive than Proctor's subsequent version of events, which understandably focused on the reaction of Catholic spokespersons, might have us believe. Some reviewers treated Tyndall's thesis with an equanimity almost approaching approbation, as in the case of both the *New York Times* and its London equivalent.[26]

The more negative responses came, predictably, from the mouthpieces of conservative elements of the religious establishment. In this context, it is relevant that the speech was delivered in Belfast, in the land of Tyndall's birth, well within virtual earshot of the many Catholic constituents of the island of Ireland. It could be seen to have been deliberately provocative.

[24] Lightman, *Victorian Popularizers*, 311. Crowe, *Extraterrestrial*, 375, also references this statement from an obituary ('Richard A. Proctor Dead' published in *The New York Times* of 13 September 1888, 1).

[25] Turner, 'Victorian Conflict', 373.

[26] The *New York Times*, 5 September 1874, 1–4, printed a review of the speech and its positive reception in the British press under the heading 'The Origin of Man'. On 7 September 1874, 3, the same journal reprinted a generally favourable review from the *London Times* entitled 'The Progress of Science'.

William Francis Barry (1849–1930), the author of a lengthy article in the *Dublin Review*, certainly thought so (albeit with a certain irony arising from the way "truths" were unreflective posited as underpinning "Faith"):

> To a Catholic, the mere account of what he holds, and the names of such as hold with him, are ample demonstration that he is really is an adversary of the truths without which Faith becomes an absurdity.[27]

The validity of Barry's assumption that the speech represented an attack on elements of Catholicism, in particular its attitude to scientific education, was confirmed by Tyndall's subsequent comments. In '*Apology for the Belfast Address*', he made it clear that he had particularly in mind a petition submitted by students at the Catholic University of Ireland in November 1873:

> It expresses the profoundest dissatisfaction with the curriculum marked out for the students of the University; setting forth the extraordinary fact that the lecture-list for the faculty of Science, published a month before they wrote, did not contain the name of a single Professor of the Physical or Natural Sciences.[28]

Thus, it would appear clear that a significant objective, if not the main motivation, for delivering the *Belfast Address* had been Tyndall's objections to the educational policies of the Irish Catholic establishment especially with respect to their lack of support for the sciences. His intention, however, was sufficiently well concealed by his overall approach to questions of science and religion and his declaration of support for scientific naturalism that it also provoked responses from other sectors of the Christian Church, including a series of lectures sponsored (and authored in part) by the Moderators of the General Assembly of the Presbyterian Church in Ireland.[29] The general tone of the comments from the liberal churches in Britain, however, was more measured, seeking to assure their

[27] William F. Barry, 'Mr. Tyndall and Contemporary Thought.' *The Dublin Review*, Oct 1876, 27, 54: 431–469 at 469.

[28] John Tyndall, 'Apology for the Belfast Address' in John Tyndall, *Fragments of Science: A series of Detached Essays, Addresses and Reviews*, vol II (New York: D. Appleton, 1892), 202–223 at 211.

[29] William Johnson (ed.), *Science and Revelation; A Series of Lectures in Reply to the Theories of Tyndall, Huxley, Darwin, Spencer, etc.* (Belfast: William Mullan, 1875).

own constituencies that science is not incompatible with revealed religion rather than rejecting Tyndall's thesis outright.[30]

It is evident therefore that the reception of Tyndall's address was divided into two broad camps. The first we might broadly summarise as those elements of the London-based press who viewed themselves as modern, intellectual and broadly liberal (in terms of religion rather than necessarily in politics). The second group, in contrast, was more conservative, generally provincial and religiously convinced or oriented. Proctor chose to mentally delete the first faction, even though his subsequent words and actions suggested that he was a fellow traveller of the metropolitan materialists, and to focus entirely on the more negative comments emanating from the proponents of revealed religion, particularly those who belonged to the Catholic Church.

Proctor's Response to Catholic Critics

Proctor privately revealed his divorce from the church to which he had adhered for only eight years in a further letter to Clodd, written in May 1875, in which he declared that he had adopted the only possible alternative to Catholic dogmatism: a position effectively akin to agnosticism—'we don't know, we won't know':

> That at such a time I was a Catholic while now I take what seems the most opposite view possible. [...] it has seemed to me that there are but two standing points—the Catholic position, or a position resembling it, on the one hand—and "ignoramus, ignorabimus" on the other.[31]

That Proctor's attitude to religion had undergone a step-change became evident to a wider audience following the publication of two articles in November 1875 in the *Chicago Daily Tribune*. In the first of these, an editorial commentary made a clear connection between Proctor's rejection of Catholicism and his espousal of "the facts of that branch of science which he so ably represents":

> Mr. Proctors' views have undergone considerable change [...] with reference to their relation to religion and revelation. Prof. Proctor was formerly

[30] See, for example, James Martineau, 'Modern Materialism: Its Attitude towards Theology', *The Contemporary Review*; 1 December 1875, 27: 323–346.

[31] Proctor to Clodd, 14 May 1875, *Clodd Correspondence*.

> a Catholic. He was recently offered a Professorship of Astronomy and Mathematics in a Catholic College in England, but found that his views were such that he could not conscientiously accept it.[32]

The second article focused on a letter to the *New York Times* in which Proctor explained his reasons for his change of position, specifically relating them to the negative reactions to the *Belfast Address*, which had forced the often dogmatic and confrontational Proctor into a pessimistic position where his "choice between these two alternatives could not be doubtful". Although his rationale is couched in quite moderate terms, it is not difficult to detect an undercurrent of anger and frustration with the representatives of his former chosen church:

> Let the dogmatic teachings of a particular Church be called A, and let the teachings of science be called B. Then either (1) A and B are consistent, or, being inconsistent, the error lies (2) with A or (3) with B. Two years ago I was satisfied with the first of these explanations. But you know what happened after the Belfast address. Theologians, who ought at least to know what their church requires, proclaimed loudly that the views B are utterly inconsistent with the views A. They convinced me that this is so; while I remained as well satisfied as ever that views B are right in the main. My obvious inference is that some error exists in views A. Theologians who teach A assure me that I must either accept A, without question, or depart from among them. My choice between these alternatives could not be doubtful. Hence simply the most marked circumstances of my change of position.[33]

Proctor did not restrict his comments on the *Belfast Address* and its consequent impact on his previous acceptance of the Catholic Church's doctrine of private judgement to this letter alone. He also made it a key point of his lecture tour in North America, adopting a presentation style in which he appeared to be delivering an assertive sermon on 'the religion of science' as again reported by the *Chicago Daily Tribune*, apparently quoting *verbatim*:

> that that fine address, his outspoken statement of the present position of science (or if you prefer it, the present position of the great majority of

[32] 'Prof. Proctor: Changed views as to Science and Religion', *Chicago Daily Tribune*, 10 November 1875, 7.

[33] 'Prof. Proctor: Germs of his present views in his earlier writings', *Chicago Daily Tribune*, 14 November 1875, 3.

> scientific men) led to an outburst of angry feeling. […] No form of vituperation was spared, and if threats of personal violence were not resorted to, they were adequately represented by warnings that an eternity of torment would reward what—even as thus misunderstood—was yet the expression of Prof. Tyndall's loyalty to the truth as he saw it.
>
> It has thus become my duty—in mere loyalty to truth—not to indicate what is my precise belief in matters religious […] but to reaffirm my views respecting the great doctrine of cosmic evolution which has become within the last few years the touchstone of scientific men.[34]

Proctor's typology of the reactions to the *Belfast Address* focused, as might have been expected given his combative stance on other issues, on the most negative commentaries which he understood to include highly intemperate language as well as the ultimate and rather more permanent sanction of the author the speech being consigned to "an eternity of torment".

One could also argue that Proctor had been driven by the combined effect of an interfaith marriage followed by academic shortfall, the loss of his son and financial collapse to adopt a religious position which was in fact significantly at odds with the milieu of scientific naturalism in which he subsequently found himself. His correspondence with Clodd on the subject of his Catholicism in 1870 had been more an indication of 'honest doubt' than it appeared at the time and was in fact a form of soul searching into which religiously informed reactions to the *Belfast Address*, and the vehement opinions of supporters of the Church of Rome in particular, acted as the final nails in the coffin rather than prime movers in the process of his unconversion. He had already moved away from traditional theology and was heading in the general direction of some form of agnosticism. As the headline attached to the previously quoted letter, published in November 1875 ('Germs of his present views in his earlier writings'), implied, it was certainly possible for a contemporary editor to perceive that Proctor's altered perspective had been some time in the making.[35]

Proctor's changing *Weltmodell* can also be read in part as a reaction to some other contemporary developments, including the publication in 1874 of *History of the Conflict between Religion and Science* by John William Draper (1811–1882). This book, aimed primarily at the "pretensions of the Roman Catholic Church and its pontiff", was the first salvo in

[34] 'Changed views', 7.

[35] 'Germs', 3.

an alleged 'war' which subsequently fed what came to be known as the 'conflict thesis'. Closer at hand, Proctor would have been well aware of the sceptical opinions of his two friends (and co-contributors to *Knowledge*) Edward Clodd and Edward Burnett Tylor (1832–1917).[36]

It is also possible that he chose to emphasise his emerging agnosticism as an element of his public science self, particularly when corresponding with journalists or addressing large audiences. The timing of his letters to the American press suggests that his combative stance may have been emphasised and publicised for the purposes of 'drumming up' larger audiences for his lecture tour. Joshua Nall has made the case that, in the years following his initial commercial success with *Other Worlds*, Proctor endeavoured to create a space for "science workers" like himself who would survive and prosper by the delivery of imaginative (read 'popular') and copiously illustrated astronomical productions, garnished with a light dusting of facts, somewhat after the style of John Pringle Nichol (1804–1859) a couple of decades earlier.[37]

In *Our Place among Infinities*, published to coincide with the American tour, he appeared to repeat the conclusions on the omnipresence and omniscience of God which had previously surfaced in *Other Worlds*. The 1875 volume, however, also expressed his newly discovered doubt and left it open to the reader to ponder the feasibility or otherwise of theological traditions. "Inconceivable that the whole universe can be for all time the scene of the operation of infinite personal power, omnipresent, all knowing." To which he added a concluding statement which expounded his new manifesto on the separation of the two epistemologies:

> Science is in the presence of the old, old mystery; the old, old questions are asked of her,—"Canst thou by searching find out God? canst thou find out the Almighty unto perfection? It is as high as heaven; what canst thou do?

[36] James R. Moore, *The Post-Darwinian Controversies: A Study of the Protestant Struggle to Come to Terms with Darwin in Great Britain and North America, 1870–1900* (Cambridge University Press, 1979), 20–29. There has in past historiography been somewhat of an overstated divergence between Victorian science and religion in the form of a 'conflict thesis'. John Hedley Brooke and others have endeavoured to remove this false dichotomy by pointing out the interconnections between theology and other epistemologies in the minds of nineteenth-century subjects. See, for example, John H. Brooke, *Science and Religion: Some Historical Perspectives* (Cambridge University Press, 1991). The views of Clodd and Tylor are discussed in more detail below.

[37] Joshua Nall, *News from Mars: Mass Media and the Forging of a New Astronomy, 1860–1910* (University of Pittsburgh Press, 2019), 25–88.

> deeper than hell; what canst thou know?" And science answers these questions as they were answered of old,—"As touching the Almighty, we cannot find Him out."[38]

The final sentence in this quotation was a pointer to the metaphysical direction of travel towards the 'Unknowable' that Proctor was to pursue during the final thirteen years of his life.

Heterodox Theological Views of Proctor and His Colleagues

Proctor's approach to religion from 1874 onwards should be read against a background of a more general move to à la carte personal theologies during the latter part of the nineteenth century.[39] He was clearly familiar with the then fashionable theories of the origins of a variety of religions which provided further support for his newly sceptical interpretation of the teachings of the church by providing an apparently rational account of how belief systems based on superstitions had become the basis of subsequent theologies. Three years before the *Belfast Address*, his colleague Edward Tylor, who was a member of the X-Club (an informal association of scientific naturalists), published his *magnum opus* entitled '*Primitive Culture*', which promulgated a whiggish hypothesis of social and religious progress. One of his distinctive contributions to the study of comparative religion, subsequently adopted by early sociologists, was the concept that animism was the earliest and most 'primitive' type of religious belief.[40] He argued that such belief systems had evolved in due course into various forms of theism: these being, in Tylor's view of comparative religion, different in content but not necessarily in type.[41]

Tylor was, of course, not the only scholar seeking explanations of the history and potential functionality of religions in societies. Contending hypotheses of religious development included the "corruption of

[38] Proctor, *Infinities*, 34. Compare with Proctor, *Other Worlds*, 333–334.

[39] José Harris, *Private Lives, Public Spirit: Britain 1870–1914* (Harmondsworth: Penguin, 1994), 169–177.

[40] Edward B. Tylor, *Primitive Culture*, 2 volumes (London: John Murray, 1871). Eric J. Carlton, *Patterns of Belief. Volume 2: Religions in Society* (London: George Allen & Unwin, 1973), 53–57.

[41] Marjorie Wheeler-Barclay, *Science of Religion in Britain, 1860–1915* (University of Virginia Press, 2010), 71–103.

language", possible Aryan origins and the deification of the sun popularised by Max Müller (1823–1900). Also available to the student of theological evolution was the competing concept of the propitiation of ancestors (sometimes in the form of 'ghosts') which was the specific viewpoint of Thomas Huxley (1825–1895) and Herbert Spencer (1820–1903). The latter idea was closely related to the parallel contemporary vogue for 'euhemerism' which sought to relate religious ideas back to real antecedent events and personalities.[42]

Another contributor to Proctor's *Knowledge* was in the process of forming his own views on the origins of religions. Clodd was a convinced evolutionist who had written biographies of Spencer and Charles Darwin (1809–1882). In 1873, he published a history of religion for children which attempted to demonstrate how humanity's thoughts on such topics as the nature of the gods had gradually evolved towards the perceived ideal of monotheism.[43] Over time, however, Clodd became increasingly agnostic. His shift was catalysed by discussions with his friends and colleagues. Lightman has observed:

> During the latter half of the 1870s Clodd became part of an intellectual circle that included prominent scientific naturalists such as Grant Allen, W. K. Clifford, T. H. Huxley, and Leslie Stephen. By 1880 Clodd could no longer accept a liberal theistic position and adopted a form of agnosticism. He believed that the idea of God, a futile attempt to explain the mysteries of existence, had been inherited from earlier, more primitive ages.

Clodd's loss of faith eventually led to the publication in 1880 of *Jesus of Nazareth*, an attempt to recover a historical Jesus as a realistic human being which included naturalistic explanations for apparently miraculous events. This book drew so much criticism from Catholic opponents, in

[42] For a further discussion of Müller's views, see Gillian Beer, '"The Death of the Sun": Victorian Solar Physics and Solar Myth.', in J. Barrie Bullen (ed.), *The Sun is God: Painting, Literature and Mythology in the Nineteenth Century* (Oxford: Clarendon Press, 1989), 159–180. See also: Wheeler-Barclay, *Religion*, 32, 37–70. Herbert Spencer, 'The origin of animal-worship, etc.', *The Fortnightly Review*, May 1870 (vol 13 old series, 7 new series): 535–550. Edward Clodd, 'The Birth and Growth of Myth', *Knowledge*, 4, 21 September 1883: 182–184.

[43] Edward Clodd, *The Childhood of the World: A Simple Account of Man in Early Times* (London: Macmillan, 1873).

particular, that Clodd subsequently "decided to avoid all direct criticism of religious beliefs in the future".[44]

Proctor must have been aware of the sceptical and rationalist views of his colleagues and, given that he published their articles in *Knowledge*, presumably approved of, or at least understood, their thinking.[45] His personal explanation of the evolution of theology was a variant of sabaeanism (also known as sabianism): the idea that primitive religions arose as a belief in the divine nature of the more prominent "heavenly orbs". Or, to put it another way, the attribution of divine characteristics to astronomical objects:

> And while we must remember how much astronomy owes to astrology, with its natural but erroneous fancies, we must not forget that in every nation of olden times, the studies of the heavenly bodies was a most important part of religion. Observation of the heavenly orbs signified observance of the gods of heaven. Last, but also highest and purest of all forms of nature worship, was that which had the dome of heaven for its temple-roof, which saw its deities visibly enthroned there or moving in solemn order above the bowed heads of their worshippers, while all earthly temples were but as altars within that glorious abode of the gods.[46]

Discussions of the 'primitive' nature of such beliefs also gave Proctor the opportunity to comment on a parallel viewpoint which he promoted in later life. Proctor held very negative views on astrology, as compared with astronomy, viewing it as dangerously close to irrational superstition or the potential entanglement of a faith-based system with 'proper science', a potential trap from which he had successfully extricated himself. Proctor accordingly wrote several articles which debunked the idea that the movements of the planets could somehow be linked to human destiny.[47]

Perhaps the best-known example of his thinking on ancient astronomy is the relatively extensive amount of effort he put into books, articles and

[44] Lightman, *Victorian Popularizers*, 256–257.

[45] See, for example, Clodd, 'Myth', 182–184 and Grant Allen, 'Our Ancestors: I – The Stone Age Men', *Knowledge*, vol 1, 24 February 1882: 351–352.

[46] Richard A. Proctor and Arthur C. Ranyard, *Old and New Astronomy* (London: Longmans, Green, 1892), 15–30.

[47] See, for example, Richard A. Proctor, 'Astrology', in Richard A. Proctor, *Myths and Marvels of Astronomy* (New York: G. P. Putnam's Sons, 1877), 1–52.

correspondence in the pages of *Knowledge*, which promoted the idea that the Great Pyramid had been constructed as an ancient Egyptian observatory, a scientific instrument, built for the purpose of making more refined astrological predictions for the sponsoring Pharaoh:

> The mystery of the Great Pyramid resides chiefly in this: that while certainly meant to be a tomb, it was obviously intended to serve as an observatory, though during the lifetime only of its builder, and was also associated with religious observances. [...] The remaining portion of the work shows how potent were those ancient superstitions about planetary influences—and their bearing first on Jewish, and later on Christian festivals and ceremonial.[48]

It seems that Proctor believed in a form of dialectic through which the empiricism of astronomy had gradually subsumed and overtaken superstition in the form of astrology, an obvious metaphor for his personal experience of the, in his view, incompatibility of religion and science.[49]

In the light of Proctor's forthright comments in 1875 on the reception of Tyndall's *Belfast Address* and his collaboration with contemporary rationalism, it is unsurprising that he abdicated from further attempts to align the relative epistemic authorities of religion and science. Nonetheless, several of his early editorial pieces in *Knowledge* were designed to reassure his potential readership by defusing a possible criticism that these two contrasting sources of meaning might be in conflict or that the study of science was somehow a path to atheism.[50]

Later in the life of his journal, he felt more able to express alternative points of view. He concurred with suggestions that his personal position was close to agnosticism and that he shared with his co-evolutionist Spencer a semi-belief in an ill-defined higher power, "The Unknowable", who was responsible for all of the elements of the universe which could not be understood by mere mortals. A position which he acknowledged was in fact barely discernible from that of agnosticism:

> In a kindly notice of my essays on "The Unknowable" in *KNOWLEDGE*, which I saw recently in *Watts's Library Guide*, the remark is made that though I "object to the term 'agnostic' as inadequately expressing my

[48] Richard A. Proctor, *The Great Pyramid: Observatory, Tomb, and Temple* (London: Chatto & Windus, 1883), v–vi.

[49] Proctor, *Infinities*, 313–323.

[50] The second article in the first edition of *Knowledge* was such an apologetic. Richard A. Proctor, 'Science and Religion. By the Editor', *Knowledge*, vol 1, 4 November 1881: 3–4.

> views," I am in sympathy with those who are recognised as belonging to the agnostic school of thought." I am not sure that I have ever expressed any objection against the term "agnostic," though I admit that no such term can quite adequately express the views of a school; but emphatically I am in sympathy with the school called, for convenience of reference, Agnostic.[51]

Lightman and Crowe have both observed that Proctor's change of religious direction can be traced to the mid-1870s and that he "was also indebted to Spencer's theistic version of agnosticism, unlike both Allen and Clodd".[52] As David Gange has pointed out: "This was a [Spencerian] metaphysic in which worship of an Absolute that transcended human conception should persist, but in which all anthropomorphism, Christology, and Church or Bible teaching must be thrown to the winds."[53] Henceforth, Proctor adopted a position which tended towards agnosticism but retained a vague theism, although statements of specific belief are noticeable by their absence from his later works. His theological 'divorce' had provided the space in which his emergent ideas about a non-Trinitarian god of nature could be overtly expressed. Despite his acceptance of the signifier 'agnostic', it is clear that Proctor, in the view of at least some contemporaries, was not per se an irreligious man. A "reverent and religious" approach to life seemed to have been a point of principle with him, as noted in one obituary:

> Another stupid slander sometimes uttered was that he had no religion, than which nothing could conceivably be more false. For theology his contempt was supernal; but no one could be more reverent and religious than he who now lies still and cold so far from his native land.[54]

Universal Evolution

Lightman has argued that Proctor's metaphysical world view was based on a conviction that the driving force behind all changes in the universe is evolutionary: whether operating at a microscopic or a macroscopic scale. Karl Guthke concurs that Proctor disavowed teleology and design in his

[51] Richard A. Proctor, 'Religion and the Unknown', *Knowledge*, vol 10, 1 July 1887: 197. Earlier essays on the topic of 'the Unknowable' can be found in 'Death of Worlds', *Knowledge*, vol 3, 4 May 1883: 263 and 'The Unknowable; or, the Religion of Science', *Knowledge*, vol 9, 1 December 1885: 37.

[52] Lightman, *Victorian Popularizers*, 315. Crowe, *extraterrestrial*, 375.

[53] David Gange, *Dialogues with the Dead: Egyptology in British Culture and Religion, 1822–1922* (Oxford University Press, 2013), 201–202.

[54] William Noble, 'Richard A. Proctor', *The Observatory*, vol 11, (1888): 366–368 at 368.

later work, taking instead a developmental view of the universe which he derived from his reading of *The unity of worlds and of nature* by Baden Powell (1796–1860), to which he added his own belief in an omnipotent and designing being.[55] The expandable scale of evolution also accounted, in Proctor's mind, for the view that there may, at the very largest scale, be clusters of galaxies like our own which contrasted strangely with his negative views on the possible extra-galactic origin of Herschel's observed nebulae.[56]

At the mid-point of his cosmic scale, Proctor perceived a role for planetary evolution. The gas giants in our solar system were to be placed at an earlier point on a hypothetical developmental timeline and had therefore not yet sufficiently condensed to the point where they might constitute solid bodies, whereas the inner planets are currently much more suited to the support of life.[57] He thus became somewhat of a 'fence-sitter' on the question of the probability of extraterrestrial life when compared with the commentators discussed in the two preceding chapters. Given these conclusions, and in the context of his developing distaste for religious enthusiasm, he proceeded to construct a halfway house between the populous cosmos of the Evangelist Brewster and a Whewellian vision of man as the only higher form of life.

In 1870, whilst professing an adherence to Catholicism, Proctor's stance on pluralism was more Brewsterish than Whewellian and therefore more in favour than not of a universe populated by a wide variety of living beings. By 1875, in response to the negative reactions of his partners in faith to Tyndall's *Belfast Address* and influenced by his own increasing scepticism, which in turn was partly founded on the ideas of his contemporaries and colleagues, he had decided to re-examine his reception of the empirical evidence for extraterrestrial life. He concluded, in partial agreement with Whewell, that the prospects were less clear than he had previously thought. Whilst accepting that the universe was shaped by the action of secondary laws including Darwinian evolution, a position clearly at

[55] Lightman, *Victorian Popularizers*, 313–314. Karl S. Guthke (tr. Helen Atkins), *The Last Frontier: Imagining Other Worlds, from the Copernican Revolution to Modern Science Fiction* (Cornell University Press, 1990), 341–343.

[56] Herschel's correspondence with Proctor on the topic of the extra-galactic nebulae (or absence thereof) was included as an addendum to Richard. A. Proctor, *Other Suns than Ours: A Series of Essays on Suns – Old, Young and Dead* (1887 repr. London: Longmans, Green, 1903), 393–419.

[57] See, for example, Proctor, *Infinities*, 67–68.

odds with the creationist beliefs of the Catholic Church, he now claimed that it was possible to divorce science from religion entirely. Science, he averred, was able to say nothing specific about a theistic deity: "so far as Science is concerned, the idea of a personal god is inconceivable".[58]

Richard Proctor thus took up two different positions on pluralism at a time when his theological position had also moved considerably, following a narrative arc which in some ways can be seen as offering a marked contrast to the often-discussed example of his sometime co-religionist and co-evolutionist, St John Jackson Mivart (1827–1900), who underwent a conversion to the purest form of Darwinism before recanting and reverting to a somewhat unbending Roman church.[59] As a consequence of the ways in which public discourses of religion and science had developed during the 1860s and 1870s, Proctor was more easily able to accept that it had become socially and psychologically feasible to clearly differentiate between these two overlapping epistemologies than had Mivart. It also seems likely that Proctor's conversion to his temporarily adopted religion, which took place over a period of time at the age of 30 or thereabouts, was rather more superficial than the deeper-rooted Catholicism of Mivart which had been generated by "a rather painful adolescent pilgrimage [...] at age sixteen".[60]

To put it another way, Proctor, unlike Whewell or Chalmers, perceived that he no longer needed to 'square the circle' of attempting to straddle two cultures whose discourses he perceived to be incommensurable, but would instead nail his colours to the mast of science whilst offering a vague reverence to the Unknowable. His colleague Edward Clodd recollected his 1870 correspondence with Proctor when commenting on his own evolving thought processes:

> A long time was to pass ere I came to see that there is no half-way house between Catholicism and Agnosticism; and that the intermediate beliefs lacked the authority which has the glamour of antiquity and the audacious assumption of finality.[61]

[58] Proctor, *Infinities*, 2.

[59] Robert J. Richards, *Darwin and the Emergence of Evolutionary Theories of Mind and Behavior*, (University of Chicago Press, 1987), 353–363.

[60] Ibid., 353.

[61] Clodd, *Memories*, 59–60.

Proctor was already in the process of adopting a thoroughly evolutionary genesis for the universe at all scales, which bore little relation to any ontology proposed by traditional theologies, when contemporary events conspired to force him to choose between his previous tacit support for the Catholic church and his emergent career as a populariser of science. A re-reading of his private and public correspondence demonstrates that it was primarily his response to some of the more febrile responses of supporters of traditional religion to a provocative and apparently materialist speech given in Belfast by John Tyndall (1820–1893) which justified his adoption of a more agnostic stance in person and in print. It was 'the straw which broke the camel's back' in terms of completing a severance from Trinitarianism, which had been some time in the making.

In a parallel but related change, he began to significantly play down the possible existence of plentiful extraterrestrial life: an idea which he had hitherto associated with the more enthusiastic and proselytising proponents of religion such as the Scottish Evangelists Chalmers and Brewster. Proctor's interior debate about religion ran parallel to a re-evaluation of the empirical evidence which did not seem to support the pluralist views of his predecessors. Shifting to a Whewellite position on this question also liberated him from having to exercise the kind of contorted thinking about natural laws which had characterised his 1870 attempt to provide naturalistic explanations for the awe-inspiring attributes of the deity.

PART III

The Nebular Hypothesis

CHAPTER 6

John Pringle Nichol, the Nebular Hypothesis and Progressive Cosmogony

This world was once a fluid haze of light,
Till toward the centre set the starry tides,
And eddied into suns, that wheeling cast
The Planets:

Alfred Lord Tennyson, '*The Princess*'[1]

On a cold but unusually clear night in December 1845, a Scottish astronomer positioned himself more than 50 feet above the ground at the mouth of a giant telescope in the midlands of Ireland. From his precarious perch, John Pringle Nichol (1804–1859) peered through "the unsteadiness in the air" at a large nebula in the constellation of Orion in an attempt to determine whether it consisted mainly of gas or was totally composed of stars. "It was—owing to the incompleteness of the instrument and unfavourable weather—the first time that the grand telescope had been directed towards the mysterious object [...] I went in breathless interest to its inspection. Not yet the veriest trace of a star!"[2] In Nichol's view, this stellar absence was an important proof of his conviction that the nebula,

[1] Alfred Lord Tennyson, 'The Princess' in Karen Hodder (ed.), *The Works of Alfred Lord Tennyson* (Ware: Wordsworth Editions, 1994), 227–303 at 241. Also quoted in: Richard A. Proctor, *Saturn and its System* (London: Longman, Green, Longman, Roberts, & Green, 1865), 201.

[2] John P. Nichol, *The Architecture of the Heavens* (London: John Parker, 1850), 113.

H. Carlton, *Cosmology and the Scientific Self in the Nineteenth Century*, https://doi.org/10.1007/978-3-031-05280-4_6

like the rest of the universe, was, is and would in the future be in a state of evolution: a process which might well extend to infinite depths in both space and time.

This form of cosmogony was opposed by conservative theologians who believed, on the contrary, that God had set the cosmos in place in its current state by means of a singular act of creation and quite possibly adhering to a timescale laid down in the first book of the Bible. They "saw fit to launch" at developmental cosmologists like Nichol "the charge of impiety, and to regard all such speculations, in sober seriousness, as rebellion against the attributes and activity of God".[3]

Nichol had previously been ordained as a minister, but had resigned from clerical orders following a period of serious health issues which prompted parallel changes of direction in both his chosen faith and his career path.[4] He became first a professional and then a popular astronomer as well as maintaining, at least for a while, his existing interests in progressive politics, journalism and economics.[5] He never returned to orthodox Christianity: a belief system to which he, in fact, may not have had a strong adherence. His 'default' religious views were probably more closely aligned to Unitarianism. In mid-career, he also, but temporarily, advocated a form of panentheism which emphasised God's intimate relationship with, and control over, the physical universe. His resignation from the post of "Rector in Montrose" and subsequent separation from the Church of Scotland not only allowed him to exercise his private judgement with respect to religion but also to pursue his interest in evolutionary cosmology.[6]

Nichol's comments on the results of his night of observing were published in 1850 at around the time he was admitted to a hydrotherapy clinic in Yorkshire where he was treated for a dependency on opiates that had

[3] John P. Nichol, 'Art. VI. State of Discovery and Speculation Concerning the Nebulae', *London and Westminster Review*, Jul 1836; 3, 2: 390–409 at 406.

[4] William A. Knight, *Memoir of John Nichol, Professor of English Literature in the University of Glasgow* (Glasgow: James Maclehose, 1896), 9. Professor Knight was the literary executor of John Nichol, the son of John Pringle Nichol. Knight had access to a collection of letters from John to his wife which he had collated in 1861 in the form of 'Leaves from my Life'. The first chapter of the Memoir consists of this journal, although its contents appear to have been redacted—see the editor's comment on p. 3.

[5] Simon Schaffer, 'The nebular hypothesis and the science of progress' in James R. Moore (ed.), *History, Humanity and Evolution: Essays for John C. Greene* (Cambridge University Press, 1989), 131–164.

[6] Knight, *Memoir*, 8.

gradually developed from about 1847 onwards.[7] His thoughts were expressed in a popular publication, one of several produced during his career, in which he clearly saw the science of astronomy and his personal concept of religion not as 'non-overlapping magisteria' but as synergistic components of his *Weltmodell*.[8] Nichol had reacted to an illness which had appeared at the time to be life-threatening with a significant shift in his espoused views on religion and an enhanced enthusiasm for a cosmogony no longer constrained by the creationary conservatism of the Presbyterian establishment.

Several historians of science have found the fluctuating fortunes of the nebular hypothesis during the nineteenth century to be a microcosm of the ways in which developments in the pattern science of astronomy have been articulated and communicated to peers and to the public. The interchanges between proponents and opponents of the hypothesis were subsequently and somewhat simplistically presented as one of many skirmishes in the hypothesised wider 'war' between science and religion. In practice, although it is a reasonable assumption that most supporters of the idea of an extended period of creation during which solid bodies cooled and condensed from diffuse clouds of heated gases were unlikely to hold literalist views of the Genesis creation story, they were nonetheless also unlikely to repudiate revealed religion *in toto* or to assume that naturalistic cosmogonies somehow obviated the existence of the deity. Scholars, including Stephen Brush, Simon Schaffer and Marilyn Ogilvie, have also pointed out the significance of the political agenda being pursued by supporters of the nebular hypothesis and the reforming objectives which underlay some of their public statements.[9]

[7] John P. Nichol, *Memorials from Ben Rhydding: Concerning the Place, Its People, Its Cures* (London: Charles Gilpin, 1852).

[8] A German word meaning 'world picture' which has previously been employed to describe both the cosmological models of physical scientists and the creation stories found in traditional religions. It thus encompasses the two key epistemologies under discussion here.

[9] Coverage of the historiography of the nebular hypothesis can be found in a number of articles including: Stephen G. Brush, 'The nebular hypothesis and the evolutionary world view', *History of Science*, 25 (1987): 245–278, Michael. A. Hoskin, 'Rosse, Robinson and the Resolution of the Nebulae', *Journal for the History of Astronomy*, 21, 4, (Nov., 1990): 331–344, Marilyn. B. Ogilvie, 'Robert Chambers and the Nebular Hypothesis', *The British Journal for the History of Science*, 8, 3 (Nov., 1975): 214–232. Schaffer, 'nebular', Simon Schaffer, 'On astronomical drawing' in Caroline A Jones and Peter Galison (eds.), *Picturing science, producing art* (New York: Routledge, 1998), 441–474 and Simon Schaffer, 'The Leviathan of Parsonstown: Literary technology and scientific representation', in Timothy Lenoir (ed.), *Inscribing Science: Scientific Texts and the Materiality of Communication* (Stanford: University Press, 1998), 182–222.

In the process of uncovering the various causes and effects of the shifting views of the selected subjects, this section will also examine some of the ways in which progressive thinkers like Nichol were able to present the 'new astronomy' to broader audiences. These public performances enabled astronomers to dovetail into parallel debates about the nature and duration of terrestrial formation processes and which supported the changing understanding of terrestrial development then being argued by geologists. New interpretations of the age and extent of the universe not only enhanced the case for the longevity of the earth but also considerably extended the eighteenth-century findings which had unseated the solar system from its formerly central position. Such concepts, like the burgeoning potential population of extraterrestrials, potentially repositioned human beings as relatively minor players who may not actually be the final cause of a vast creation.

Origins of the Nebular Hypothesis

The idea, proselytised by Nichol, that the current components of the cosmos had begun their existence in "a fluid haze of light" was not, as we have noted, universally accepted. As might be expected, the preferred outcomes in this ontological debate had at least as much dependence on the theological views of the participants as they did on astronomical observations. In a manner analogous to the pluralism discourses which were examined in the preceding section, the proponents and opponents of nebular cosmogony entangled their views of stellar evolution with their personal theologies.

The debate itself revolved around conflicting explanations of the origins of the stars and planets, that is, the cosmogony of the universe itself. Traditional explanations, based on the biblical narrative, held that God had created the cosmos in one short, relatively recent and presumably highly energetic burst of activity and that the various heavenly bodies had pursued stable courses in response to divine fiat during the several thousand years which had elapsed since the genesis of the world.[10] In due course, however, some astronomers came to believe that the traces of the

[10] Sara J. Schechner, *Time & Time Again: How Science and Culture Shape the Past, the Present and Future* (Cambridge MA: Collection of Historical Scientific Instruments, Harvard University, 2014), 14–15.

developmental processes which led to the birth of stars and planets were still visible in the form of the gaseous material contained in the strange bodies known as 'nebulae' and that these processes might still be ongoing in the universe at large.

The first idea which subsequently became a component of the broader nebular hypothesis sought specifically to explain the origins of the solar system. It predated the nineteenth century, there having been several earlier proposals which postulated the evolution of our planetary system from a nebular cloud, including the writings of Emmanuel Swedenborg (1688–1772), Thomas Wright of Durham (1711–1786), Johann Heinrich Lambert (1728–1777) and Immanuel Kant (1724–1804).[11] The version of this concept most frequently referred to by later writers was the attempt by Kant to provide an explanation for how the various bodies then known to exist in the solar system had come to be formed. In *Universal Natural History and Theory of the Heavens* (1755), he conjectured that in a slightly irregular primeval universe composed mainly of gas, gravity alone could account for the formation of the solar system.[12]

Pierre-Simon Laplace (1749–1827) built on the views of his predecessor by pointing out that the shared orbital directions and similar inclinations to the ecliptic of most of the objects in our solar system is highly improbable on the conservative hypothesis of a random distribution, but fits well with the idea of condensation from a rotating nebula, thus neatly side-stepping the violation of the principle of conservation of angular momentum which had been an inadvertent by-product of Kant's proposal.[13] Laplace conveniently chose to almost overlook the apparently retrograde rotation of the satellites of Uranus which was to prove a thorn in the flesh of several subsequent supporters of his hypothesis.[14]

[11] Ogilvie, 'Chambers', 216. Michael J. Crowe, *The Extraterrestrial Life Debate 1750–1900: The Idea of a Plurality of Worlds from Kant to Lowell* (Mineola NY: Dover Publications, 1999), 97–98, asserts that a number of astronomers dispute the idea that Swedenborg had described an early form of the nebular hypothesis.

[12] Gerald J. Whitrow, 'The role of time in cosmology', in Wolfgang Yourgrau & Allen D. Breck (eds.), Cosmology, *History and Theology* (New York: Plenum Press, 1977), 159–177 at 167.

[13] Ibid., 167–168.

[14] Stanley Jaki, *Planets and Planetarians: A History of Theories of the Origin of Planetary Systems* (Edinburgh: Scottish Academic Press, 1978), 123–124. The English version of Laplace's nebular hypothesis appeared in Pierre S. Laplace, (tr. J. Pond) *The System of the World* (London: Richard Phillips, 1809) Volume II, Book V, Chapter VI, 354–375. Laplace omitted Uranus from his list of known satellite rotations, on page 357.

A second nebular hypothesis came about as a result of observations made by Frederick William Herschel (1738–1822), who spent a considerable period of time classifying the various nebulae that could be observed with his telescopes in a fashion somewhat analogous to the classifications of living things which were then in vogue amongst naturalists.[15] Herschel, having studied many phenomena which could loosely be categorised under the general heading of nebulae, gradually came to the conclusion that those which were unresolvable (at least by means of the technology at his disposal) are formed from a "shining fluid". These objects he believed were future sidereal systems in their nascent stage.[16]

Some commentators tended to combine these parallel accounts of the origins of stars and planetary systems into a single overall concept when making their interventions. The potential for such confusion was assisted by the combination of the two ideas under the newly minted title of the "nebular hypothesis" in a relatively brief chapter contained in the third *Bridgewater Treatise*, written by William Whewell (1794–1866). As will become evident in the narrative of nebular observations in Ireland in Chap. 7, this semantic ambiguity could be intentionally deployed to conflate a disproof of one form of the nebular hypothesis with the other.

Whewell's 1833 *Bridgewater Treatise* not only combined two different arguments into an apparently unitary hypothesis but also set the stage for the significant contribution made by Nichol. Whewell, a liberal Anglican, at that time acknowledged the value of natural theology and accepted its central tenet that the existence and beauty of the universe are evidence of the handiwork of the deity.[17] It may seem somewhat surprising, therefore, that he took a quasi-empiricist view of the nebular hypotheses of Herschel and Laplace and held that they might turn out to be valid narratives of cosmic development, although they were awaiting confirmatory evidence.

Whewell's suspicion that nebular cosmogonies might be solidified as a result of future developments turned out to be an almost self-fulling prophecy. In 1853, he was, as discussed in Chap. 4, eventually forced into

[15] Simon Schaffer, 'Herschel in Bedlam: Natural History and Stellar Astronomy', *The British Journal for the History of Science*, 13, 3 (Nov., 1980): 211–239.

[16] Stephen G. Brush, 'The nebular hypothesis and the evolutionary world view', *History of Science*, 25 (1987): 245–278, at 251–252.

[17] William Whewell, *Astronomy and General Physics Considered with Reference to Natural Theology* (London: William Pickering, 1833), 181–191.

a position of fairly uncritical acceptance of true nebulosity as one leg of support for his version of the functioning of final causes. At that point, he decided to accept that the existence of unresolvable nebulae had been confirmed by the southern hemisphere observations of the Magellanic Clouds by John Frederick William Herschel (1792–1871), thus simultaneously, although inadvertently, opening the door to the possibility that nebular evolution is a real phenomenon. It is evident that he was compelled to adopt this slightly awkward posture because it suited his thesis that human beings are a unique life form, singularly destined to be saved by a unique revelation. Whewell's' *Bridgewater Treatise* had fortuitously anticipated such an outcome by claiming that it would not at all detract from his overarching theme of God as "a wise and good Creator", adding a moral dimension as well as a purpose to the construction of the universe:

> Leaving then to other persons and to future ages to decide upon the scientific merits of the nebular hypothesis, we conceive that the final fate of this opinion can not, in sound reason, affect at all the view which we have been endeavouring to illustrate; the view of the universe as the work of a wise and good Creator.[18]

In contrast with Whewell's initially cautious approach, Nichol was one of the most vocal popularisers of the hypothesis in both of its forms. Schaffer's view is that Nichol was the prime mover who created and popularised a discourse of nebular cosmogony from a few strands of available narrative.[19] Nichol did indeed play a significant role in bringing the idea to public notice, although it is also evident that preceding authors had published various earlier forms of the hypothesis and that it had already been utilised in heated (and heat-based) arguments about geological formation processes during the 1820s.[20] Nichol published a brief review of the ideas of Laplace and Herschel in the same year as Whewell's *Bridgewater Treatise*, but it was not until 1836 that he wrote a lengthier essay, published in the *London and Westminster Review*, in which he expounded an extended exploration of his evolutionary ideas.[21]

[18] Whewell, *Astronomy*, 190–191.

[19] Schaffer, 'nebular', 134–135.

[20] Ogilvie, 'Chambers'. Philip Lawrence, 'Heaven and Earth – the relation of the nebular hypothesis to geology', in Yourgrau and Breck (eds.), Cosmology, 253–281.

[21] John P. Nichol, 'On some recent discoveries in astronomy', *Tait's Edinburgh magazine*, 4.19, October 1833: 57–64. Nichol, 'State of Discovery'.

John Pringle Nichol's Early Career in Polemics

As a relatively young man, Nichol is reported to have been precociously successful in achieving academic recognition: "at college he got every prize he ever tried for." He also suffered from unspecified health problems and a penchant for flights of imagination:

> There are still people at Hawick [where Nichol was for a time headmaster of the Grammar School] who can tell stories of the thin handsome youth who astonished every one by his brilliancy, and interested all the ladies by his weak health and poetic fancies.[22]

The positive feedback accorded to his "weak health and poetic fancies" seems to have provided him with useful resources for later public performances and the production of popular scientific literature during his subsequent career.

It is also clear that by the time of his appointment to the chair of Astronomy at the University of Glasgow in 1836, Nichol shared with some of his peers a central project of social reform.[23] While he was "still prosecuting his studies for the Church of Scotland", he accepted "the position as the editor of the *Fife Herald*", where he horrified "the landed interest of Fife by revolutionary articles in the Herald; While Robert [his brother] and he talked, walked and speculated together on the poetry of science".[24] Nichol's political views were of such a nature that it is easy to understand that some of his future opponents in the nebular hypothesis debates were in no small part also reacting to what they perceived to be the potentially dangerous views of "'those washed & unwashed Radicals' who had kept the country in a state of alarm since the Reform Bill."[25] They viewed the ideas of 'progressive' cosmology as being in some way bound up with an agenda derived in large part from Ricardian political economics and promulgated by the 'Philosophic Radicals'.

Such a consonance of epistemologies could also be found in the articles which Nichol subsequently contributed to several liberal publications

[22] Knight, *Memoir*, 6.

[23] Schaffer, 'nebular', 142–143.

[24] Knight, *Memoir*, 6–7.

[25] James A. Secord, *Victorian Sensation: The Extraordinary Publication, Reception, and Secret Authorship of Vestiges of the Natural History of Creation* (University of Chicago Press, 2000), 227.

including *Tait's Edinburgh Magazine* and the *London Review*.[26] Amongst the range of topics covered by Nichol in these articles written during a period of time, 1832 to 1837, which saw both the passing of the first Reform Act and considerable opposition to the fixing of grain prices, were tracts on such themes as the inequity of the Corn Laws, the injustice of the use of tithes to support the Church of Ireland and the repressive approach of some members of the protestant ascendancy in that country with regard to both the economic and spiritual concerns of their tenants:

> Let him murmur against the tax, and, if he be a tenant-at-will, a *notice to quit* will be the answer. If he have a lease, its renewal will depend upon the readiness with which he pays his tithes; and, even while the term of the lease may be remote, the landlord has numerous modes of chastising him, should he dare to lift up his ideas to the prospect of religious equality.[27]

His view on the use of tithes to support the established church in a predominantly Catholic country also aligned with his subsequent opposition to the enforcement of a particular theological viewpoint on would-be educators:

> Moved by anxiety that a religious spirit shall pervade all teaching,—[...]— the Founders of many of our Educational Institutions (among others, of the Scottish Universities) have sought to secure fitting dispositions in the Instructor, by demanding that, previous to his induction in office, he subscribe the special articles of a Church.[28]

One of his most prominent early opponents, at least in Nichol's own mind, was Thomas Chalmers (1780–1847). Nichol had taken specific aim at his fellow Scot on the topic of Chalmers' 1832 attack on those political economists who denied the influence of Malthusian principles. In his

[26] Schaffer, 'nebular', 142–147. NB the *London Review* was renamed *London and Westminster Review* in 1836.

[27] John P. Nichol, 'The Irish Tithe Bill', *Tait's Edinburgh magazine*; Jul 1837; 4, 43: 402–404 at 403.

[28] J. Willm, *The Education of the People: A Practical Treatise on the Realm of Extending Its Sphere & Improving Its Character. With a Preliminary Dissertation on Some Points Connected with the Present Position of Education in This Country by J. P. Nichol, LL. D.* (Glasgow: William Lang, 1847), lxii. As noted in Chap. 2, this opposition to 'subscription' was one of the few viewpoints which Nichol would have shared with the natural theologian William Paley (1743–1805).

rebuttal, Nichol employed the rhetorical device of almost-excessive praise followed by the dismantling of one's opponent's argument which features, for example, in Shakespeare's narrative of Mark Anthony's funeral oration for Julius Caesar.[29] In Nichol's view, Chalmers not only believed in a cosmological steady state in which God rules over an array of stellar systems which were created in their final, stable forms but had also projected similar principles of stability and lack of progress onto "the only tolerable" political system, an absolutist monarch ruling a feudal society:

> Dr. Chalmers possesses too much prudence to admit of his entirely cutting himself loose; but he has never taken sufficient note of the progress of events; and his sentiments on all practical matters are marked, accordingly, by a tendency towards absolutism and mere fancifulness. There is an amusing *naiveté* in that announcement somewhere in the volume referred to, of the form of government which he considers to be the only tolerable one. He wishes a King upon the Throne, and nobles around him, clad, we fancy, in mail jackets, and swelling with the *virtues* and *charities* of feudalism.[30]

Nichol's Breakdown

Nichol had, despite his later reservations about the role of the established Church in oppressing Irish tenants and the standardisation of theology within education, "passed through the Divinity Hall [of The University and King's College of Aberdeen], and was, before attaining his majority, licensed as a preacher" by the Church of Scotland, but "he only appeared two or three times in the pulpit, and at one of these [...] he completely broke down. Whatever was the cause, he soon abandoned theological study, and devoted himself to that of astronomy."[31]

It is clear from this comment that two important changes in Nichol's *Weltmodell* occurred almost simultaneously: a break with the Church of Scotland and a significantly increased interest in the science of astronomy.

[29] Thomas Chalmers, *On Political Economy in Connection with the Moral State and Moral Prospects of Society* (Glasgow: William Collins, 1832). John P. Nichol, 'Dr. Chalmers', *Tait's Edinburgh magazine*, Nov 1832; 2, 8: 189–195. William Shakespeare, 'The Tragedy of Julius Caesar', Act 3, Scene 1.

[30] Nichol, 'Dr. Chalmers', 194.

[31] 'Professor Nichol', *The Gentleman's Magazine: And Historical Review*, Nov 1859; 207: 536.

One contemporary biographer, the Glasgow publisher James MacLehose (1811–1885), described the events which led to Nichol moving away from the public avocation of Presbyterian orthodoxy in the following terms:

> a change in his theological views led, shortly after his ordination, to his leaving the ministry, and giving himself up to the educational work to which, in one form or another, his life was devoted [...] In 1834 temporary ill health led to his resignation".[32]

A more immediate account of Nichol's illness can be found in one of the relatively few extant letters written either to or by him.[33] His publisher, John Stuart Mill (1806–1873), writing to him in December 1834, began his epistle with the following comments:

> I am grieved to hear of your narrow escape, and most heartily congratulate you and myself that the danger is past. By all means keep yourself as quiet as possible, and do not even think of any intellectual exertion till you are completely recovered. I once lost a most valued friend, one of the most valued I ever had—though not to be compared with you in intellect—in consequence of a similar disease—the eldest son of Tooke, the political economist. I believe he brought on the malady almost entirely by intense and unremitting study. I most earnestly entreat you, for the sake of us all, not to expose yourself to a similar danger.[34]

This version of events makes slightly strange reading in that the friend who Mill mentions, Eyton Tooke (1806–1830), actually took his own life. He had reportedly "been severely unwell with a severe cold since Christmas, and his spirits had been much depressed" and had also "frequently complained of pains in his head". "A verdict of Insanity was returned."[35] It is very difficult to know to what extent the story of Tooke's "similar disease" and suicide is to be read as directly comparable to Nichol's case. Mill's concern that his author and friend should avoid "any intellectual exertion"

[32] James MacLehose, *Memoirs and Portraits of One Hundred Glasgow Men* (Glasgow: James MacLehose & Sons, 1886), 249.

[33] Knight, *Memoir*, xiii.

[34] Frances E. Mineka (ed.), *The Earlier Letters of John Stuart Mill 1812–1848 (Volume XII of the Collected Works of John Stuart Mill)* (University of Toronto Press, 1963), 245, Letter 117, to John Pringle Nichol, dated 18 December 1834.

[35] 'Suicide of Eyton Tooke, Esq.', *The Observer*, 31 January 1830, 3.

must also be tempered by his subsequent assumption that Nichol could complete a promised article for the *London Review* by the end of January, which doesn't seem to allow for a lengthy recovery period. In any case, we would be justified in concluding that, in Mills' view, Nichol's illness had complex origins which were in some way connected with his state of mind as well as his physical well-being.

John Nichol (1833–1894), who was in due course appointed Regius Professor of English Literature at the University of Glasgow, provided a further account of this period in his father's life. He reported, at second or third hand, that his father's change of heart with respect to religion coincided with a recurrence of a "rapid fever". According to Nichol junior, there had been a real concern that his father might not survive his latest illness:

> My father had again fallen into bad health, and had to resign his position as Rector in Montrose. Old people whom I saw there last winter told me stories of him, and how few expected his life would last, he looked so thin and worn. The same illness, and circumstances connected with it, brought about a change in his religious views; his mind was turned away from the Church of Scotland, after he had been for some years a licentiate and preached several sermons, and became more and more exclusively directed into the paths of Literature and Science.[36]

Nichol's change of religious perspective in this version of events seems to have had a direct connection with his preceding illness.[37] This summary is of significant value in any attempt to understand Nichol's shifting *Weltmodell* and appears to some extent to run parallel to the experiences of Thomas Chalmers, but with a significantly different, indeed opposing, narrative arc: from public ecclesiastical conformity towards private doubt. We should, however, note that Nichol junior was too young at the time for this version to be based on his own memories: he was relaying what his father and others had told him.

It is difficult to retrospectively assemble a consistent narrative of the events reported by John Nichol, MacLehose and others due to some contradictions in timing. Although MacLehose asserted that John Pringle Nichol had resigned due to ill-health in 1834, according to *The Scotsman* he delivered lectures in Edinburgh in the guise of the "Rev. J. P. Nichol of

[36] Knight, *Memoir*, 8–9.
[37] MacLehose, *Memoirs*, 249.

Montrose" during April 1835. From the same journal, we also learn that in or before October that year "friends to Mr. Nichol, and admirers of his talents, gave him a dinner in the Waterloo Hotel, as a Welcome to him, on his establishment in this city".[38] So it seems that his actual resignation, following an illness in late 1834, actually took place at some point in the summer of 1835. It remains the case that both his biographer and his son were convinced that this resignation was the result of "a change in his religious views".

Unlike the case of Chalmers, we unfortunately have limited evidence as to the state of Nichol's *Weltmodell* before, during and after his distinct change of theological direction. It is possible, but obviously not demonstrable, that the dissonance between the theological views that he was expected to extol in his clerical role and his private religious inclinations formed a component of the evidently complex disruption of his brain–body system which led to his breakdown. After his medical and religious crisis, Nichol never returned to the fold of the Church of Scotland. In 1841, for example, objections were raised to his possible candidacy for the post of Dean of Faculties at the University of Glasgow on the grounds that "they [Nichol and a fellow academic] had not produced evidence of having signed the Confession of Faith".[39]

It is probable in fact that he only ever had a semi-detached adherence to the Calvinist orthodoxy of the Church of Scotland, or as his son put it succinctly: "In religion he thought for himself, and, of course, arrived at his own conclusions."[40] The extensive library in Nichol's house included the writings of a number of eighteenth-century Unitarian divines.[41] Nichol's general approval of Unitarianism and, in particular, of the lectures given by the American Unitarian theologian William Ellery Channing (1780–1842) is further confirmed by an anecdote dated to the early 1850s.[42] Towards the end of his own life, the younger Nichol again mentioned his father's shared religious preference: "'I am a Christian,' Nichol

[38] 'Astronomy', *The Scotsman*, 28 March 1835, 1. 'Mr J. P. Nichol.', *The Scotsman*, 24 October 1835, 3.

[39] James Coutts, *A History of the University of Glasgow; From Its Foundation in 1451 to 1909* (Glasgow: James MacLehose and Sons, 1909), 419.

[40] Knight, *Memoir*, 51.

[41] 'Professor Nichol', *The Scottish Review*, January 1860: 1–17 at 12.

[42] Knight, *Memoir*, 118–121. The reference to "Dr. Channing" is attributed by Nichol junior to a "Rev. A. Mackennal".

[the son] said to me in one of our last talks; and he went on to add that he was a Unitarian, and that his father had been the same."[43]

Given that he had been raised with some affinity for the tenets of Unitarianism, Nichol would have had difficulty in accepting the doctrines that only predestined 'saints' would or could be saved.[44] The key difference between Nichol and Chalmers, therefore, and the probable explanation for their divergent responses to severe illness lies in Nichol's preceding exposure to religious heterodoxy. Under stress, he instinctively reverted to a position which was no longer compatible with a calling to the ministry in the Church of Scotland.

Professor of Astronomy at the University of Glasgow

Nichol's post-unconversion career was mainly focused on astronomy, a discipline which, in the 1830s, was not considered to be a requirement for any profession and lay "outside the course prescribed for the Degree in Arts":

> In 1836 he was appointed, by the Crown, Regius Professor of Astronomy in the University of Glasgow—an election which has some additional interest from the fact that Carlyle was among the competitors. The subject of astronomy being outside the course prescribed for the Degree in Arts, and none of the professions requiring any knowledge of it, the number regularly enrolled in the class was and has been very limited.[45]

Fortunately, given the relatively sparse enrolment in his classes, Nichol was not confined to a single subject or faculty but like his contemporaries in academia could claim an implicit authority to extend his sphere of operations well outside his original remit. With the tacit support of his peers, he engaged, or offered to engage, in the teaching of natural philosophy, natural history and moral philosophy.[46]

[43] Ibid., 301. This comment is also attributed to Mackennal, but the conversation seems to belong to a much later period, that is, 1891–1892.

[44] The University of Aberdeen has been associated over time with a level of sympathy for the Anglican Church and latitudinarian views. See, for example, Ned C. Landsman, *From Colonials to Provincials, American Thought and Culture 1680–1760* (Cornell University Press, 1997), 64.

[45] MacLehose, *Memoirs*, 249.

[46] Coutts, *History*, 388.

The pending appointment was nonetheless of sufficient public interest, presumably because of the way in which the new astronomy was seen to be entangled with theology, to be widely covered in the press. Given also Nichol's lack of experience in the subject and his provocative position as a political commentator who had been engaged in writing polemics for radical journals, the contested space in which the role of chair in astronomy came to be awarded was clear to both the newspapers and to Nichol's supporters who attempted to enlist the powerful aid of a former opponent:

> Radicals such as the MP and fellow Scot Joseph Hume [...] were powerful allies in Nichol's application for the Glasgow astronomy chair when it became vacant in January 1836. The Liberal press campaigned for Nichol, Tory newspapers protested, and help was prudently sought from the great evangelical leader Thomas Chalmers.[47]

The possible involvement of Chalmers in Nichol's campaign seems surprising given their apparently differing political and theological views. On the other hand, it is perhaps of relevance that they did in fact share some similarities of background. Both were relatively well-educated middle-class men who originated from adjacent counties in rural eastern Scotland. The more overtly progressive of the two men, Nichol, however, was born 24 years later than the Presbyterian Evangelist, so had little or no recollection of the discourses surrounding the French Revolution which seemed to have been instrumental in blunting the nascent egalitarian tendencies of Chalmers and his contemporaries.[48]

The potentially problematical request for assistance from Chalmers also serves to remind us that the older man was not in fact a stereotypical Tory in terms of his political thinking:

> He opposed Parliamentary Reform but supported Catholic Emancipation, repeal of the Corn and Combination Laws, reduction of tariffs, and an income tax, while his closest political associates—Wilberforce, Calthorpe, and Whitmore—all supported the Liberal Tories.[49]

[47] Schaffer, 'nebular', 149.

[48] Stewart J. Brown, *Thomas Chalmers and the Godly Commonwealth in Scotland* (Oxford University Press, 1982), 28–29.

[49] Boyd Hilton, *The Age of* Atonement*: The Influence of* Evangelicalism *on Social and Economic Thought 1785–1865* (Oxford University Press, 2001), 61.

Chalmers also held positive views on the utility of education which were similar to those of Nichol, and there was perhaps more convergence between the two men on questions of religion than one might initially assume.[50] It is also worthy of note that Chalmers had been one of Nichol's sponsors for the post of Rector of Montrose Academy, albeit at a time, some eight years earlier, when Nichol was known for his precocious academic excellence and unusually wide subject knowledge rather than for the Radical polemics which began to appear in the early 1830s. "He obtained the highest prize in the University of Aberdeen" and "was a teacher at thirteen years of age, mathematical master in Cupar Academy at twenty".[51]

Nichol, an existing member of academic staff "who had shown great aptitude and versatility as a lecturer" would have been seen by the university authorities as clearly preferable to the controversialist and metropolitan views of Thomas Carlyle (1795–1881), a polymath and essayist who was living in London whilst working on a history of the French Revolution.[52] In a similar contest held in 1854, for example, Nichol's son, whose *Weltmodell* incorporated a similar tendency to youthful radicalism, was 1 of 60 students who unsuccessfully supported the candidacy of Carlyle as Rector of the University of Glasgow, despite the risk of being "publicly held up as 'Deists', 'Atheists' and 'Pantheists.'"[53]

Carlyle was significantly less qualified than Nichol for the job which he appeared to have applied for more in hope than expectation during a period of enforced impecunity whilst writing his as-yet-unpublished three-volume magnum opus. According to one biographer:

> In a moment of discouragement at the beginning of 1836, he even wrote to propose himself for a professorship of astronomy which had become available at Glasgow. Fortunately, nothing came of it.[54]

Nichol's Exposition of the Nebular Hypothesis

Nichol's appointment to the chair in astronomy was followed by the publication of a major article on the nebular hypothesis. In this paper, Nichol argued that the cosmos had unfolded gradually over time in response to rules and processes which may have been stochastic in their effects and

[50] Brown, *Chalmers*, 114–115, 191–192.

[51] 'Mr J. P. Nichol', *The Scotsman*, 24 October 1835, 3. Schaffer, 'nebular', 145.

[52] Coutts, *History*, 388. John Nichol, *Thomas Carlyle* (London: Macmillan, 1892), 63–73.

[53] Knight, *Memoir*, 114.

[54] Fred Kaplan, *Thomas Carlyle: a biography* (Cambridge University Press, 1983), 232.

therefore, or so it would appear to some opponents of the hypothesis, not under the control of a deity. He was particularly exercised by earlier objections to the hypothesis which had been raised by some theologians. William Paley (1743–1805), for example, had dismissed Laplace's speculation as being one of "the guesses of those who reject an intelligent Creator".[55] Chalmers, though happy to populate other planets with living beings, had claimed in his *Bridgewater Treatise* that:

> The dispositions of matter in the planetary system were fixed at the original setting up of the machine. The laws of matter were ordained for the working of the machine. The former, that is, the dispositions, make up the framework, or what may be termed the apparatus of the system. The latter, that is, the laws, uphold the performance of it.[56]

In his determination to follow an independent and progressive academic stance, Nichol unsurprisingly failed to agree with his fellow Scot on the applicability of traditional arguments based on the initial creation of all matter in its final form, subsequently regulated by the predictable effects of the divinely ordained laws of motion as described by Newton.[57] He dismissed the view that the supposed instantaneous formation and collocation of matter in stellar systems acted as evidence for the activity of a creator, arguing that we would not accept that the gradual and observable processes by which leaves "are evolved from the almost unorganized germ" would provide a good case for the existence of God. Nichol complained in a footnote that "certain writers [Chalmers] have strangely thought fit to launch at the nebular hypothesis the charge of impiety" by a "common and gross abuse of the argument from final causes".[58] Nichol's views on collocation and the nebular hypothesis continued to be debated well into the 1840s as the "university reformers" attempted to "simultaneously combat the threats from both biblical literalism on the one extreme and aggressive materialism or naturalism on the other".[59]

[55] William Paley, *Paley's Natural Theology*, vol II (London: Charles Knight, 1836), 18.

[56] Thomas Chalmers, *On the Power, Wisdom and Goodness of God: As Manifested in the Adaptation of External Nature to the Moral and Intellectual Constitution of Man* (1833 repr. Glasgow: William Collins, 1839), 30–31.

[57] Hilton, *Atonement*, 363.

[58] Nichol, 'State of Discovery', 406–408.

[59] Crosbie Smith, *The Science of Energy: A Cultural History of Energy Physics in Victorian* Britain (London: Athlone, 1998), 25–26.

Nichol concurred with Whewell both on the operation of secondary laws as the underlying 'machinery' of cosmogony and on the dangers of proofs based on current ignorance of the development processes operating in nature. He perceived the latter as being potentially falsifiable claims which, if anything, would lend support to materialist arguments rather than mitigating against them. Unlike Whewell, however, he did not hold to the notion that God must have set up the initial conditions of the universe such that the outcome of the operation of secondary laws in support of final causes must inevitably follow. Nichol believed that evolutionary processes derived from and mediated by natural laws had been and still were operative in the universe, but it is unclear both from these early writings and from his subsequent productions as to what extent his underlying and almost pantheistic "Infinite Essence" was disposed to intervene from time to time to guide terrestrial and celestial events.[60] Given his progressive political viewpoint, he also needed and wanted to allow room for the human race to achieve future improvements by dint of its own capabilities and efforts.

Nichol's Political Views Post 1836

Although the events of 1836 represented a clear movement away from any possibility of a career in the established church to an avocation in the field of astronomy, Nichol's overt political radicalism, although now unconstrained by any religious affiliation, seems to have slowly diminished thereafter, as evidenced by a cessation in the production of polemical writings from 1837 onwards. This alteration of his views on politics well have been influenced by the structural changes which accompanied his appointment to a senior role in academia. The general tenor of his subsequent output offered the merest nod in the direction of Trinitarian theology and very little political commentary other than continued support for the benefits of an education unconstrained by religious orthodoxy and unsuccessful attempts to promote some of his more progressive colleagues into academic posts. His potential power base within the University of Glasgow was, however, significantly constrained by his lack of progress in creating a new observatory despite his over-enthusiastic spending on equipment:

[60] John P. Nichol, *The Stellar Universe: Views of Its Arrangements, Motions, and Evolutions* (Edinburgh: John Johnstone, 1848), vii–viii. Also see comments on his engagement with occasionalism below.

"Nichol was broke and his finances under attack from the conservative Faculty."[61]

In due course, he appeared to undergo a slow drift towards a Liberal-Conservative position, thus again aligning him more closely with Chalmers than might have been predicted during the early years of his career. MacLehose's opinion was that Nichol's views in relation to politics and religion pursued opposing directions of travel during his lifetime:

> In his early years a Radical, as the term was then understood, he latterly inclined to what has been called Liberal-Conservativism, his ideal of an English statesman being more nearly represented by Lord Palmerston than by either Mr. Gladstone or Mr. Disraeli. Inspired, as is everywhere apparent in his works, by a deep spirit of reverence, he was in his later days more and more sceptical as to definite doctrines.[62]

In parallel with MacLehose's summary of the political stance adopted by Nichol, his son noted: "In politics he was a temperate radical, verging, about the close of his life, towards becoming a whig." Nichol junior, however, also noted some rather more nuanced ideological positions as he reported his father's continuing interest in radical politics in other European countries during the 1850s, including meetings and correspondence with the nationalist revolutionaries Giuseppe Mazzini (1805–1872) and Lajos Kossuth (1802–1894). Nichol's second marriage, in 1853, to the abolitionist and suffragist Elizabeth Pease (1807–1897), motivated him to add to his political portfolio the cause of emancipation in the United States: "the great Anti-Slavery movement of those days".[63]

Nichol and Continuous Creation

For a period in the 1840s, Nichol's personal version of metaphysics underwent a temporary aberration, becoming akin to that of a Malebranchian occasionalist. Some of his peers believed that this continued to be his position for the remainder of his life. After his death, for example, one obituary described him as having held God to be an ever-present entity who, like a Platonic demiurge, "literally creates the universe every moment"

[61] Schaffer,' nebular', 149–150. For Nichol's views on education and religion, see Willm, *Education*, xi–lxxxiii.

[62] MacLehose, *Memoirs*, 252.

[63] Knight, *Memoir*, 100–121, 58.

rather than being a chronologically distant deist progenitor.[64] Nichol's views were echoed in typically more florid fashion by his friend George Gilfillan (1813–1878):

> Nichol [...] never loses sight of the universe as an instant, ever-rushing emanation of the Deity. "God," he says, quoting a friend of kindred spirit, "literally creates the universe every moment." He is led by Boscovich's theory of atoms to suppose an infinite Will, producing incessantly all force and motion.[65]

These two accounts of Nichol's ontological beliefs do indeed seem to describe a more monist approach, though there is little or no evidence from his later writings to suggest that he continued to support such a view. These comments from an obituarist and a popular essayist seem to indicate some distance between their authors' understanding of Nichol's views and his own later statements on the topic of religion, perhaps indicating that, as in the complex case of his political proclivities, Nichol was particularly adept at presenting a variety of selves which seemed to be designed to fit the various different contexts in which he operated and the disparate audiences with whom he interacted.[66]

Both sources cited above have in fact reproduced elements of a paragraph which originally appeared in Nichol's 1846 book *Thoughts on Some Important Points Relating to the System of the World*, where this panentheist concept was attributed to "a valued friend yet destined to hold, if he is spared in life and health, one of the most distinguished places in the science and literature of the country". The complete original quotation reads:

> God literally creates the Universe every moment. Every instant is a new morning of creation. He is the Alpha and the Omega; the beginning and the ending; the Creator, the Sustainer, the Provider, 'in whom we live, move, and have our being.[67]

[64] 'Professor Nichol', 5–6.

[65] George Gilfillan, *A Second Gallery of Literary Portraits* (Edinburgh: James Hogg, 1850), 247.

[66] See discussion in Chap. 8 of the ideas presented in *The Stellar Universe.*

[67] John P. Nichol, *Thoughts on Some Important Points Relating to the System of the World* (Edinburgh: William Tait, 1846), viii. The original quotation was subsequently reproduced in a collection of Brown's lectures: Samuel Brown, 'The History of Science (1846)', *Lectures on the Atomic Theory and Essays Scientific and Literary.* vol 1 (Edinburgh: Thomas Constable, 1858), 299–344 at 332.

Nichol had taken this quotation verbatim from a lecture given earlier in the year by the "valued friend" Samuel Brown (1817–1856), who was mainly noted for his popularisation of the views of a Ragusan Jesuit, Roger Joseph Boscovich (1711–1787), on the identical nature of the atoms of different substances which, if true, would lead to the possibility of transmutation of elements, thus symbolically reawakening the exoteric objectives of some preceding alchemists. Brown's 1843 application for the Chemistry chair at the University of Edinburgh was rejected due to the failure of other chemists to reproduce his self-reported success in transmuting carbon into silicon. In 1846, Brown was continuing to lecture on the topic of chemistry and also disseminated his view that a fourth phase of the history of society should be added to the three previously proposed by Auguste Comte (1794–1859): a stage in which faith would have priority over reason as the driver of development and discovery. This line of argument had at that time also impressed itself on such luminaries as "Sir William Hamilton, [Francis] Jeffrey, [Thomas] Chalmers, [Julius Charles] Hare, [Thomas] Carlyle, [Sir Robert] Christison, and [Edward] Forbes".[68]

There seems little doubt, therefore, that Brown's lecture had, at least for the time being, influenced Nichol while he was in the process of preparing *Thoughts*. Nichol's occasionalist flirtation can also be seen as foreshadowing the later attempt by the Jesuit Gerard Manley Hopkins (1844–1889) to reconcile the finite life of the physical world with the hoped-for continuity of God's sustaining presence and the preservation of the soul.[69] This novel metaphysics potentially solved the thorny question of how divine control can be exerted in an apparently chaotic universe, even though Brown's thesis in many ways ran contrary to Nichol's previous position on secondary laws and seemed to limit the possibility of autonomous and progressive human development. Nonetheless, this excursus into an exotic alternative provides us with a further glimpse of the plasticity of Nichol's *Weltmodell* and his continuing partiality to novel and potentially disruptive concepts, including a parallel and well-documented flirtation with phrenology.[70]

[68] Spencer T. Hall, *Biographical Sketches of Remarkable People, Chiefly from Personal Recollection, Miscellaneous Papers and Poems* (London: Simpkin, Marshall, 1873), 54.

[69] Niamh Brown, 'Devotional Cosmology: Poetry, Thermodynamics and Popular Astronomy, 1839–1889', PhD thesis submitted to the University of Glasgow (2016), 45–62.

[70] Schaffer, 'nebular', 148.

Nichol's uncritical acceptance of Brown's unconventional views on chemistry seems in practice to have been a brief deviation during the mid-1840s. The idea of constant intervention does not sit well with the vaguely imagined universal architect found in most of Nichol's preceding and subsequent works. Excluding the Brownian episode, his religious beliefs evolved over time after his rejection of a theological vocation to a sceptical position on the topic of revealed religion, although his public pronouncements were somewhat constrained both by the controversies which accompanied the publication of *Vestiges of the Natural History of Creation* and by his own bankruptcy.[71]

By 1848, Nichol's interventionist deity had reverted to an unchanging and uninvolved "Power" who had originally orchestrated creation but appeared to be no longer intimately involved in the evolution of the unreal or insubstantial universe inhabited by humans:

> Or below, is there not a Power, fixed, abiding—an energy which changes in feature only, (as the expression of one's countenance,) but which itself never changes—a grand substratum which alone is true existence—the only positive reality in the Universe?[72]

These rather transcendental reflections do not suggest that a coherent theology was operating in Nichol's mind: more a vague intuition that the visible cosmos is in fact a transitory illusion. He also did not seem to express a belief in a personal connection with a theistic deity. His 'Power' does not intervene directly in his, or anyone's, personal life and is thus an entity not much better defined or more approachable than the vague conceptions of the unknowable god of nature subsequently invoked by Richard Anthony Proctor (1837–1888) and Herbert Spencer (1820–1903).[73]

Progressive Evolution

In 1837, Nichol reiterated and expanded his version of the nebular hypothesis in a book entitled *Architecture of the Heavens*, which took the form of "a series of letters to a lady". The subtitle can be read as a

[71] Secord, *Sensation*, 466–467, documents Nichol's negative reaction to the publication of *Vestiges* and provides a reference to the announcement of his bankruptcy in *The Times* of 30 March 1842.

[72] Nichol, *Stellar*, vii–viii, 18.

[73] See Chap. 5.

periphrasis implying that the work was suitable for a less well-educated and less-affluent audience. Its object was "to popularize the stupendous wonders that have been discovered in astronomy, [...] during the present century".[74] One of these wonders, in his view, was the vast distances from earth at which stellar objects might now be presumed to lie. At one point in the book, he inferred a distance to the furthest resolvable star clusters (as observed through Herschel senior's largest telescope and as previously estimated by Herschel himself) of around two million light years.[75]

One contemporary reviewer reinterpreted Nichol's statements in "this admirable work" to the effect that the universe is infinite in both duration and extent. This reading led the reviewer to recognise that under these circumstances time could become transcendental in an eternal universe: "the reader will have impressed upon him the infinity of space; and the eternity of time, if that can be called time which hints no idea of a beginning, no prospect of an end."[76] But, as we shall see in Chap. 8, although Nichol sought on several occasions to inflate the universe well beyond this 1837 estimate, his opinions on the nature of time and space, like many of his other thoughts, were somewhat fluid and his works do not exhibit a consistent idea of whether or not the universe is finite.

It has already been noted that the notion of developmental processes for stellar and solar systems aligned well with Nichol's earlier radical views on the political and social progress of humanity. The nebular hypothesis, which remained one of his key themes throughout his popular astronomy lectures, was also a key ingredient of a master narrative of evolution at all levels: the other main components being geological and biological development theories. All three of these related concepts seemed to require

[74] John P. Nichol, *Views of the Architecture of the Heavens: In a Series of Letters to a Lady* (Edinburgh: William Tait, 1837). 'Dr. Nichol's Architecture of the heavens (Book Review)', *The Spectator*, 22 July 1837; 10, 473: 689–690.

[75] William Herschel, 'Catalogue of 500 new nebulae', *Philosophical Transactions of the Royal Society*, 1802, 92.2: 477–528. Quoted in S. G. Brush, *The Nebulous Earth: The Origin of the Solar System and the Core of the Earth from Laplace to Jeffreys* (Cambridge University Press, 1996), 35. Nichol, *Architecture*, 37–46.

[76] The *Spectator* reviewer quoted Nichol's estimate of the distance to the farthest visible objects to be 11,765,475,948,678,678,679 miles. This translates to approximately 2,000,000 light years. Corrected for the current value of the distance to Sirius (which Nichol used as his standard of brightness), the contemporary solution would be 2,580,000 light years. 'Architecture', *Spectator*, 690. The reviewer's final phrase is an unattributed (but presumably well-known) quotation from James Hutton, *Theory of the earth, with proofs and illustrations. In four parts.* vol 1 (Edinburgh: William Creech, 1795), 200.

long timescales for their processes to unfold. His adoption of nebular cosmogony assisted at least two elements of this overall argument by enabling him to successfully 'bridge the gap' and to bring in parallel ideas of lengthy chronologies which supported the second leg of the tripod, namely the sphere of geology.

Nichol's subtext of combining astronomical and geological evolutionary processes and timescales is clearly indicated in *Architecture* by the way in which he endeavoured to extend the concept of the lengthy development of the earth's surface to our planetary neighbours:

> See now the supposed geological aspect of our fellow planets and the cessation of the riotous disturbing volcanic powers as they get old.[77]

This coincidence of developmental ideas was one of several concepts picked up and expanded in *Vestiges*.[78] The authorship of this controversial synthesis of developmental hypotheses was variously attributed to several prominent Victorians, including Nichol himself, but was eventually revealed to be the Scottish publisher Robert Chambers (1802–1871).[79] The Chambers brothers (Robert and William) were the proprietors of *Chambers' Edinburgh Review*, a liberal organ which had already printed, and would continue to print, a number of favourable editorial comments on Nichol's works.[80]

Chambers was conscious of the probable objections which he would encounter from the conservative voices of those religious enthusiasts whom he referred to as "the saints", to the extent that he added a terminal disclaimer to his work in an attempt to deflect future criticism and censure, or in his terms: "some views about religion which will help greatly, I think, to keep the book on tolerable terms with the public, without compromising any important doctrine".[81] Nonetheless, the negative reactions which greeted the publication of *Vestiges*, a book which bore more than a passing

[77] 'Architecture', *Spectator*, 690.

[78] See Chap. 4. Paul Elliott, 'Erasmus Darwin, Herbert Spencer, and the Origins of the Evolutionary Worldview in British Provincial Scientific Culture', 1770–1850', *Isis*, 94, 1 (March 2003): 1–29.

[79] James A. Secord, 'Behind the veil: Robert Chambers and *Vestiges*' in Moore (ed.), *History*, 165–194, at 186. Secord, *Sensation*, 495–498.

[80] See, for example, 'Professor Nichol's Views of the Architecture of the Heavens.', *Chambers' Edinburgh journal*, 29 July 1837: 210–211.

[81] R. Chambers, Letter to A. Ireland (1844) in Secord, 'Behind the veil', 171–172.

resemblance to his own publications, must have been rather worrying for Nichol at a time which coincided with a personal financial crisis. This convergence of events would account for his preference to avoid being conflated with either authorship of the book or promotion of its ideas.[82]

Following his severe illness and the theological breakdown of 1834–1835, he had repurposed himself as a promotor of both professional and popular astronomy, having put particular faith in a theory of cosmogony which obviated the need for a 'once and for all' creation of the universe in its final form. His belief in the nebular hypothesis was, however, about to be put to the test, as we shall see in Chap. 7, by a series of apparently empirical observations made by the very instrument he had helped to construct.

Nichol represented a relatively new phenomenon in academia, having become a professional astronomer after the serious illness which led to his 'divorce' from the established church. As with the case of Thomas Chalmers, his emerging interest in and deployment of astronomical topics enabled him to combine the language and habitus of one emotional environment, the established church, and deliver them into the parallel worlds of academia and popular science with considerable success. At the same time, he began to tone down his previously declared support for radicalism and political progress, although he retained an interest in the progress of nationalist revolutions and the secularisation of education throughout the remainder of his life.

[82] Coutts, *History*, 388–389.

CHAPTER 7

"And Eddied into Suns, that Wheeling Cast/ The Planets": Rosse, Robinson and the Leviathan of Parsonstown

> But the lower lip, which is drawn inwards with the curve of a conch shell,—oh what a convolute of cruelty and revenge is there! Cruelty!—to whom? Revenge!—for what? Ask not, whisper not. Look upwards to other mysteries. In the very region of his temples, driving itself downwards into his cruel brain, and breaking the continuity of his diadem, is a horrid chasm, a ravine, a shaft, that many centuries would not traverse.
> Thomas De Quincey, 'System of the heavens as revealed by Lord Rosse's telescopes'[1]

Observations made in Ireland in the mid-1840s which had the objective of demonstrating that a totemic astronomical object, Messier 42 (more commonly known as the Orion nebula), could be resolved into discrete stars were employed by conservative theologians and their supporters in an attempt to demolish the nebular hypothesis and its associated narrative of extended timescales and post-creation developmental changes within the universe. The eventual outcome of the production of this apparently empirical evidence was that, even though it appeared at the time to have settled the debate decisively in favour of a wholly stellar basis for nebulae in general, the Orion observations did not have the effect of destroying evolutionary cosmogonies to the extent that some observers and commentators might have hoped for and expected. Despite the way in which

[1] Thomas De Quincey, 'System of the heavens as revealed by Lord Rosse's telescopes', *Tait's Edinburgh magazine*, September 1846; 13, 153: 566–579 at 571.

H. Carlton, *Cosmology and the Scientific Self in the Nineteenth Century*, https://doi.org/10.1007/978-3-031-05280-4_7

aristocratic authority had been asserted via control of the means by which truth claims were tested and the management of the transmission of results, doubts eventually crept into the minds of both the astronomers and their audiences.

The two main actors to be considered in some detail in this chapter are the amateur astronomer William Parsons (1800–1867) and his co-observer Thomas Romney Robinson (1792–1882). Parsons inherited his father's title and estate in 1841, becoming the third Earl of Rosse, an Anglo-Irish peer and a prominent member of the protestant ascendancy. Robinson was a Church of Ireland minister and director of the Armagh Observatory. Rosse and Robinson both held structural positions in Anglo-Irish society, as a landlord and semi-professional astronomer respectively, sometimes set them in political and theological opposition to the majority of the inhabitants of the island, some of whom were required to fund their aristocratic and astronomical activities via rents and tithes.

Prompted by apprehensions arising from the disruptive potential of the metastable Irish context, they were, to differing degrees, inclined to deny the concept of nebular evolution.[2] Rosse's motivation for studying nebulae also seems to have been bound up in a sense of filial duty which was motivated by the impact on his father of the loss of William's younger brother at a relatively young age. He successfully combined his intention of carrying on his father's interest in the interaction between astronomy and theology, culminating in the construction of a giant telescope, with an emergent enthusiasm for engineering projects leading to his family's subsequent engagements with technological innovations and industrial enterprises.[3]

The observations of various nebulae previously identified as probably being gaseous in nature provides a further illustration of the ways in which

[2] Jim A. Bennett, *Church, State and Astronomy in Ireland* (Belfast: Armagh Observatory, 1990), 59–84, 104–108. See also Simon Schaffer, 'The nebular hypothesis and the science of progress' in J. R. Moore (ed.), *History, Humanity and Evolution: Essays for John C. Greene* (Cambridge University Press, 1989), 131–164, and the discussion of Nichol's political views below.

[3] Seventh Earl of Rosse, 'William Parsons, third Earl of Rosse', *Hermathena*, No. 107 (Autumn 1968): 5–13, at 12, "The second son, Clere, became a highly successful railway engineer and was responsible for constructing railway lines throughout the world, especially in South America; while the youngest, Charles, became world-famous as the inventor of the steam-turbine."

the subject's *Weltmodell* drives epistemological outcomes.[4] In the case of the nebular hypothesis, the management of both social and instrumental interactions was particularly evident in the way in which Robinson used telescopic observations in support of his theological and political opposition to nebulosity. He was able to bolster his arguments by eliciting the support of the difficult-to-challenge authority of his patron.

The Leviathan of Parsonstown

The 'Leviathan of Parsonstown' was a six-foot reflecting telescope built by Rosse at Birr Castle in Parsonstown (now the town of Birr in County Offaly) in the midlands of Ireland. This was actually a relatively unsuitable location in view of the restricted opportunities for observation it provided due to the prevailing climate.[5] The "great telescope" also had limited manoeuvrability, being supported by two great stonework pillars which significantly constrained its lateral movement[6]:

> Astronomers spend weeks at Parsonstown, and yet perhaps enjoy only one or two really good nights for observation. When to these atmospheric impediments is added the comparatively limited lateral range of the great telescope, a long space of time will appear to be required for making a through and general search of the heavens.[7]

Despite these obstacles, in early 1846 Parsons succeeded in training the Leviathan on the Orion nebula, perhaps the most notable of a list of nebulae previously reported to be unresolvable by Frederick William Herschel

[4] A German word meaning 'world picture' which has previously been employed to describe both the cosmological models of physical scientists and the creation stories found in traditional religions. It thus encompasses the two key epistemologies under discussion here.

[5] Simon Schaffer, 'Astronomers mark time: Discipline and the personal equation', *Science in Context*, 2, 1, (March 1988): 115–145 at 126. Parsonstown subsequently reverted to the more traditional name 'Birr', and King's County is now known as County Offaly. Michael. A. Hoskin, 'Rosse, Robinson and the Resolution of the Nebulae', *Journal for the History of Astronomy*, 21, 4, (Nov., 1990): 331–344 at 333, 341.

[6] Thomas R. Robinson, 'On Lord Rosse's Telescope (Continued)', *Proceedings of the Royal Irish Academy (1836–1869)*, vol 3 (1844–1847): 114–133 at 118–123.

[7] 'Travels in Search of Lord Rosse's Telescope', *Chambers' Edinburgh Journal*, 26 December 1846; 156: 401–404 at 404.

(1738–1822), and had found that it apparently could be resolved into individual stars and therefore did not consist of Herschel's 'shining fluid'.[8]

The decision to build the world's largest telescope in such an evidently unsuitable location obviously raises the question of Rosse's motivations for carrying out such a difficult project, given also that the end product would be of limited usefulness. This question has been studied by Trevor Weekes, who concluded that Rosse's father, Laurence Parsons (1758–1841, second Earl of Rosse), was the most likely key influence in his son's future interest, noting that William's early education took place at Birr Castle under the supervision of his parents.[9]

Parsons junior subsequently graduated from Trinity College Dublin and then spent one year at the University of Oxford where he was awarded first-class honours in mathematics. He joined the Royal Astronomical Society in 1822. "By 1827, he had begun experiments in mirror manufacture", but his work on more sizeable telescopes did not really take off until after his marriage in 1836 to a wealthy Yorkshire heiress, Mary Field (1813–1885), who was able to provide additional funding for her spouse's enthusiasm for this particular form of engineering project.[10] By 1840, he was collaborating with Robinson, a fellow astronomer, who became his main cheerleader as well as being a key factor in his motivation for his project of constructing large telescopes and the search for true nebulosity.[11]

Laurence Parsons and the Death of John Clere Parsons

Prior to his marriage, there had been one other significant event which influenced the future Lord Rosse's decision to engage with entangled questions of cosmology and theology. The relatively early death of his younger brother John Clere Parsons (1802–1828), the only one of Laurence's five children not to outlive him, gravely affected his then

[8] Thomas R. Robinson, 'On the Present Condition of the Earl of Rosse's Great Telescope', *Proceedings of the Royal Irish Academy (1836–1869)*, vol 4 (1847–1850): 119–128.

[9] Trevor C. Weekes, 'Origin of the 3rd Earl's interest in astronomy' in Charles Mollan (ed.), *William Parsons 3rd Earl of Rosse: Astronomy and the Castle in Nineteenth-Century Ireland* (Manchester University Press, 2014), 30–43.

[10] Weekes, 'Origin', 34. Alison, Countess of Rosse, 'Part 2: Family Life at Birr' in Mollan (ed.), *Parsons*, 55–70. Seventh Earl of Rosse, 'William Parsons'.

[11] Bennett, *Church*, 104.

elderly father. An ensuing "long period of deep affliction" led in due course to the publication of the second Earl of Rosse's slightly unexpected but thoughtful treatise, *An Argument to Prove the Truth of the Christian Revelation*, which he dedicated to his late son:

> He died on the 10th of August, 1828, when it was fondly hoped that he was just about to gather the fruits of his honourable labour. During the long period of deep affliction for so great a loss, I studied the subject of this argument.[12]

The main theme of this book was an argument, which Laurence developed at some length, to the effect that Moses had correctly described the sequence of events which unfolded during the creation of the world. Since the Patriarch could not have had first-hand knowledge of this history, he must therefore have received a divine revelation, thus establishing both the truth of religion and the accuracy of Genesis. A contemporary reviewer of the book noted the affective connection between "the hope of meeting the lost one [his son] in another and a better world" and Parsons' endeavour to "apply the evidence of external nature directly to the proof of Revelation, without the intervention of Natural Religion".[13] This hope was dependent on a more theistic interpretation of the creation than was currently being espoused by the deistic processes favoured by some natural theologians. It was thus not difficult for readers to appreciate that Parsons senior's desire to prove the correctness of revealed religion was strongly linked to the premature death of his second son and was in due course to lead to a performance of filial duty by his eldest son which will be seen in Chap. 8 to be similar to the case of the Herschel family.

In the course of his argument the second Earl attempted to unravel the origins of planet earth. Laurence included some favourable references to the nebular hypothesis of "La Place", albeit perhaps without a complete understanding of the implications thereof. He also demonstrated some familiarity with current astronomical thinking, including a reference to the *Treatise on Astronomy* recently published by John Frederick William Herschel (1792–1871) and expressed a hope that future scientific discoveries would provide further support for his thesis: "in the science of

[12] Second Earl of Rosse, *An Argument to Prove the Truth of the Christian Revelation* (London: John Murray, 1834), iii.

[13] 'Argument to prove the truth of the Christian revelation', *The Athenaeum*, 26 April 1834; 339: 311–312.

astronomy, that much remains to be discovered? [...] And also that many discoveries will be made respecting those luminous appearances in the heavens, called nebulae?"[14] Laurence Parsons' grief-driven thesis, with specific reference to the question of the origins of the mysterious nebulae, provided a motive for his son both to undertake the construction of the Leviathan and to set the agenda for the initial observations to be made with it.

It is also clear that both Laurence and his eldest son were located in a complex and sometimes dangerous theological and political space which in due course provided additional reasons to pursue his telescope construction projects. The second Earl had been an early supporter of Catholic emancipation to the extent of donating land for the building of a Catholic parish church well before the passing of the Roman Catholic Relief Act.[15] From the mid-1820s to the early 1840s, however, the parish of Birr was the location of one of the few nineteenth-century Catholic schisms in Ireland: an occasionally violent breakaway proto-protestant movement led by the Crotty cousins, Michael (1794/5–1862) and William (1808–1856), "who became respectively, a Church of Ireland parson and a Presbyterian minister after having [...] first attempted to organise a dissident group of Catholics at Parsonstown".[16]

That the effects of the rift created by the Crottys continued until at least October 1845 can be seen in the comments made by a Scottish Evangelist, James Carlile (1784–1854), who had been despatched to Birr on a proselytising mission by the Presbyterian Synod. He eventually acceded to local Catholic opinion, in the form of letters written to the press by the parish priest Father John Spain (1796?–1848), and agreed to withdraw from his attempts to address the local Mechanics Institute "because Mr. Spain and I cannot agree on certain points of theology such as the right and duty of all men to read the Scriptures, and the efficacy of masses for the dead".[17]

[14] Second Earl of Rosse, *Argument*, 13–14, 40–41, 77–78, 411–416. John F. W. Herschel, *A Treatise on Astronomy* (London: Longman, Reese, Orme, Brown, Green & Longman, 1833) is mentioned in a footnote on page 34.

[15] Seventh Earl of Rosse, 'William Parsons'.

[16] K. Theodore Hoppen, *Ireland Since 1800: Conflict and Conformity* (London: Longman, 1999), 79.

[17] Margaret Hogan, 'William Parsons' influence on the town and community of Birr' in Mollan, (ed.), *Parsons*, 90–121 at 96–99. 'Printed address by J. Carlile to the Roman Catholic members of the Mechanics Institute of Parsonstown', 25 October1845, Birr Castle MSS J 9.

Another indication of theological perturbances in the parish was the 1846 report, subsequently played down, of an attack on the Leviathan itself: "I suppose you have heard of the insane attempt which has been made to destroy Lord Rosse's great reflector upon Religious Grounds."[18] This allegation, even if untrue, suggests that there was significant local antipathy to a perhaps poorly understood and potentially hubristic device which one commentator opined might have been attacked "on the ground that it looks too profanely into the mysteries of creation!" It seems that the linkage between astronomy and religion was also very much present in the minds of the writers and readers of *The Athenaeum*, even if vicariously projected into the mind of a local "mischievous fanatic".[19] The alleged attack took place at the height of the potato famine so it is not difficult to alternatively adduce a more political motive for animosity towards this symbol of the ascendancy.[20]

Political Upheavals and the Irish Famine

King's County (later County Offaly) and the surrounding area were embroiled in several political controversies during the early nineteenth century, including the rise in 'agricultural' crimes and other upheavals leading up to the passing of the 1832 Reform Act. Such was the enthusiasm for this measure amongst his mainly Catholic constituents that William Parsons, a Conservative who at that time held the title of Lord Oxmantown, felt impelled to appear to support the Whig government of Lord Grey (1764–1845).[21] In due course, he also had to contend with several powerful movements amongst the local population who had initially pursued the cause of Catholic Emancipation and subsequently the repeal of the Act of Union. The first of these initiatives had looked to the nationalist Daniel O'Connell (1775–1847) as its figurehead and reached such peaks of

[18] Humphrey Lloyd to John Herschel, 17 October 1846, Royal Society Herschel MSS HS 11.285. Emphasis in the original.

[19] The supposed attack was also reported in the weekly gossip column of *The Athenaeum*, dated 17 October 1846, 1072. This article, however, also suggested that the original report may have been spurious.

[20] Hogan, 'William Parsons', 103–111.

[21] Andrew Shields, 'The public life and political opinions of the 3rd Earl of Rosse' in Mollan (ed.), *Parsons*, 122–158 at 133.

unrest that in one instance William and his father felt it necessary to call out the local militia.[22]

The sequence of events which seems to have most perturbed William was a series of attacks on landlords and their agents, which culminated with the murder, in January 1839, of his relative by marriage and near neighbour Lord Norbury (1781–1839) "in broad daylight, just outside the walls of his home at Durrow Abbey near Tullamore".[23] The violent death of his friend occurred during the construction of William's first major telescope, a three-foot reflector which was completed in late 1839. The death of his neighbour and the general feeling of unrest in the area contributed to William's subsequent determination to provide paid employment for some of his tenants in the early to mid-1840s, some of which he personally supervised rather than becoming a member of the growing class of absentee landlords.[24]

In contrast with these 'good works', William, now Lord Rosse, in the midst of the ongoing depredations caused by the potato famine "refused to be involved in the scheme of Outdoor Relief in late 1847, after the soup kitchens were closed down". This course of action favoured ratepayers over the poorer tenants, forcing some of the latter to seek help in the already crowded workhouse. Rosse's retrogressive decision followed a trip to London during which his "only daughter and eldest child", Alice (1839–1847), died.[25] The loss of young Alice was, according to the *King's County Chronicle*, keenly felt by the family and the local community:

> We are sure that all classes here will sympathise with the noble Earl and his amiable Countess, in this most painful bereavement. Lady Alice was a child who endeared herself to all by her kind and affectionate manner and gentle disposition.[26]

This fulsome eulogy is reminiscent of the sentiments by which Darwin appeared to be actuated following the death of his daughter Annie. It also reminds us that "deaths of children were a constant spectre in the lives of Victorian households", and, as a consequence, there were ample precedents for the forms of words which were felt to be appropriate following such a tragedy. Alice was the first, but not the last, of the Rosse's children

[22] Hogan, 'William Parsons', 94.
[23] Ibid., 112–113. Shields, 'public life', 144–145.
[24] Hogan, 'William Parsons', 104–106.
[25] Hogan, 'William Parsons', 109–111.
[26] *King's County Chronicle*, 4 August 1847, quoted in Countess of Rosse, 'Family Life', 57.

to die prematurely so it is not unreasonable to suppose that this particular death was more keenly felt than subsequent bereavements. In due course, Mary Countess of Rosse gave birth to 11 children of whom only 4 boys survived until adulthood.[27]

During the visit to the capital, which saw the death of young Alice, Rosse had also been exposed to the wave of 'famine fatigue' which followed a short but significant financial crisis. Whilst it is evident that either of these events could have had a bearing on his subsequent resistance to providing assistance in his home demesne, it seems probable that the sense of anomie which he encountered in London's business community had a significant bearing on his negative attitude to outdoor relief, which would not in any way have been lightened by the affect of the death of his daughter. "Whatever the reasons, Rosse's genial, resourceful and helpful attitude changed progressively to a sterner, stricter and more entrenched one after he returned to Birr."[28]

Unfortunately, Lord Rosse's papers and correspondence provide little direct indication of his views on such theologically contentious questions as to whether the universe is an evolving entity or how firmly he personally adhered to the tenets of the Church of Ireland. To the extent that religion, science, politics and nationality were irretrievably intertwined in his local time and space, however, it is evident that the various conflicts and privations affecting his tenants were contributory factors in Rosse's decision to undertake telescope building projects which also provided him with a diversion from quotidian angst, casting him back in time to an early interest in celestial objects which had been further fostered by his father and which had preceded his assumption of more terrestrial duties.

Rosse, Robinson and Early Observations of the Nebulae

That the initial objective for the new instrument was to refute the existence of true nebulae clearly ran counter to the second Earl's earlier provisional endorsement of Laplace's theory. One key reason can be adduced for this change in astronomical objectives. It had become evident by the 1840s that the nebular hypothesis was seen as a supporting argument for the evolutionary, progressive and materialist ideas associated with radicals

[27] Ibid., 57–58.
[28] Hogan, 'William Parsons', 109.

like John Pringle Nichol (1804–1859) and John Stuart Mill (1806–1873). It was therefore anathematic to many members of the Church of Ireland as well as to his Catholic parishioners.[29] More specifically, Rosse was encouraged and supported in his endeavours by Robinson, an enthusiastic cleric, who was strongly opposed to the, from his perspective, negative implications of the nebular hypothesis.

Robinson was a graduate of Trinity College and a Church of Ireland minister who had the living of Enniskillen and the patronage of the Primate of All Ireland. He shared several characteristics with Nichol, including having been appointed to a role as the director of the Armagh Observatory in 1823 despite having no previous experience in astronomy, but he held opposing views on both politics and religion. Robinson was, like Nichol, a powerful speaker much given to flights of hyperbolic overstatement when addressing audiences of amateur enthusiasts, making full use of the lively poetic imagination which he had developed at an early age, having been acclaimed as the "infant bard of Belfast". His upbringing in Ulster and an occupational dependency on the established church had combined to make him "an aggressive Ulster Episcopalian". One evident consequence of the heterogenous political and enthusiastic religious environments in which he was raised was that he viewed the nebular hypothesis as a barely concealed attack on the established church and so he "preached in Dublin against Papism, materialism, unbelief, and evolution theory".[30]

Robinson was apparently so convinced of the rightness of his position and so keen to undermine the concept of ongoing cosmogonic development processes that as early as November 1840 he knew what the outcome of future investigations of nebulae would be. Following a series of observations made at the castle by means of a three-foot reflector, a precursor of the Leviathan, he expressed himself in very confident terms to an

[29] See, for example, Wolfgang Steinicke, *Observing and Cataloguing Nebulae and Star Clusters: From Herschel to Dreyer's New General Catalogue* (Cambridge University Press, 2010), 104–109.

[30] Bennett, *Church*, 59–84. Lord George John de la Poel Beresford (1773–1862), who appointed Robinson to the Armagh Observatory post, became Archbishop of Armagh and therefore Primate of All Ireland in June 1822. Steinicke, *Observing*, 105. Simon Schaffer, 'The Leviathan of Parsonstown: Literary technology and scientific representation', in Timothy Lenoir (ed.), *Inscribing Science: Scientific Texts and the Materiality of Communication* (Stanford: University Press, 1998), 182–222 at 207.

audience at the Royal Irish Academy, as reported by his colleague Sir James South (1785–1867) to Lord Rosse:

> You cannot imagine the impression that his manner and matter made on the Academy, and his assurances that all Nebula observations must be reformed, brought in with uses by allusions to cosmogonic theories of certain astronomers, in which he gave vent to philippics against Herschel and the French astronomers of no ordinary sharpness drew continued cheers from his hearers.[31]

Evidently Robinson, like Nichol, was able to bring his previous experience of preaching to bear in order to intensify the way in which "his manner and matter" impressed his audience. The "philippics against Herschel", however, are somewhat ironic in view of Robinson's later appeals for assistance to the authority of the far more established astronomer. Some of the more dramatic views expressed by the cleric at the meeting were clearly at variance with those of South who had a prior agreement with Robinson "that he would not make an extempore harangue but that he would bridle his warm heart's feeling with having before him a written sermon". Despite this precaution, South reported that nonetheless Robinson "went further than I could join him".[32] Robinson's more forthright assertions were not recorded in the official proceedings of the Academy as South was subsequently able to persuade him to tone down or remove some of the pejorative comments which had been made at the actual meeting.[33]

Rosse, whilst publicly more cautious than Robinson about claims of nebular resolvability, when conversing at dinner paraphrased Robinson's claims, telling his visitors that "42 of the 47 nebulae [from a list provided by John Herschel] have been resolved" and that "he altogether rejects the nebular hypothesis".[34] We should note here that in making statements of this nature Rosse was following Robinson's lead in obscuring the plastic boundary between the two versions of nebular cosmogony. Based on the

[31] South to Rosse, 11 November 1840, Birr Castle MSS K 1.2.

[32] Ibid.

[33] Thomas. R. Robinson, 'Account of the Three-feet Telescope' in Charles Parsons (ed.), *The Scientific Papers of William Parsons, Third Earl of Rosse 1800–1867* (Cambridge University Press, 2011), 14–19.

[34] See, for example, Hoskin, 'Rosse Robinson', 338 and Schaffer, 'Leviathan', 214. Frances Power Cobbe to Louise Cobbe, dated to the 1840s, Birr Castle MSS J 7. Also see Frances P. Cobbe, *Life of Frances Power Cobbe by herself*, vol 1 (Boston: Houghton, Mifflin, 1895), 177.

use of the Birr telescopes to 'resolve' the Herschels' nebulae, they were implicitly claiming to have simultaneously undermined the Laplacian speculation on which the author of *Vestiges of the Natural History of Creation* (1844) had depended for the first leg of its controversial developmental tripod.[35]

Whilst Rosse's reported comments support the idea that refutation of the nebular hypothesis had formed a significant component of his motivation for building the telescope, he also had some wider objectives. The telescope construction projects which he undertook required him to solve some complex technical and managerial problems. For example, he needed to find a method to safely cast and accurately polish a six-foot mirror without, as discussed below, the benefit of explicit information on the processes used by William Herschel. Robert Stawell Ball (1840–1913), Rosse's last observing assistant, suggested that

> it was more the mechanical processes incidental to the making of the telescope which engaged his interest than the actual observations with the telescope when it was completed [and his] special interest in the great telescope ceased when the last nail had been driven into it.[36]

Success, then, possibly at the time and certainly *post facto*, was defined at least in part by completion of the telescope which came to be seen as a major engineering feat and an "Irish triumph" which provided ample opportunity for Rosse to associate with and impress both interested visitors to Birr Castle and the gentlemen of science in London.[37]

It is also important to note that the initial Leviathan observations took place just before the Great Potato Famine. This tragic event significantly reduced the use of the telescope between its completion in 1845 and the end of the famine in 1848.[38] It has been inferred, as noted above, that Rosse's additional objectives in the whole enterprise of the Leviathan

[35] Schaffer, 'nebular', 138–139. See Chap. 6 for a discussion of the two versions of the nebular hypothesis then in circulation and the controversy created by the publication of *Vestiges*.

[36] Robert S. Ball, *Great Astronomers* (London: Isbister, 1895), 287 quoted in Wolfgang Steinicke, 'The M51 mystery: Lord Rosse, Robinson, South and the discovery of spiral structure in 1845', *Journal of Astronomical History and Heritage*, 15, 1 (2012): 19–29 at 22.

[37] Simon Schaffer, 'On astronomical drawing' in Caroline A Jones and Peter Galison (eds.), *Picturing Science, Producing Art* (New York: Routledge, 1998), 441–474, at 456–461.

[38] Seventh Earl of Ross, 'William Parsons', 9–10.

included a need not only to demonstrate his mastery of the requisite engineering techniques but also to exercise benevolence as a landlord at a time of severe food shortages and great social dislocation.[39] Rosse had in any case provided a worked example of the capability of the Anglo-Irish landowning class and their tenants in rivalling, or indeed improving upon, the astronomical products of bourgeois Georgian England.

The initial usage of the Leviathan consisted almost entirely of observations of nebulae, as might be expected. Successful resolution of the Orion nebula appeared to be claimed by Rosse, in far more restrained and slightly ambiguous terms than Robinson's earlier dramatic delivery in Dublin, in a letter to Nichol in early 1846. This claim was subsequently reported in editions of *The Times* and *The Athenaeum*.[40] These reports were not supported by any graphical representations at the time, and, surprisingly, it was not until 1867 that a drawing of the nebula, as seen through the Leviathan, was made available to the public press. This delay reflects the great difficulty which observers encountered when attempting to capture an image of an object of interest from their slightly perilous position at the mouth of the great telescope often in conditions of limited 'seeing' and at times having to battle with optical defects which arose frequently as a result of the tarnished state of the six-foot mirror. To these uncertainties we should add the subjective difficulty of deciding whether or not one is observing a field of very fine stars scattered like "grains of sand" or a nebulous haze. It was perhaps to be expected that detailed illustrations of nebulae were few and far between.[41]

Mechanical Objectivity and the Art of Drawing

This dearth of illustrations also highlights the more general difficulties which confronted early Victorian astronomers and publishers when endeavouring to present a reliable image of a view as seen through a telescope. Several stages of inscription and translation occurred in this chain of events. Starlight travelled through the unpredictable Irish skies via a large but perishable mirror to the eye of an observer. The observer in turn was required to alternate between looking through the eyepiece and attempting to reproduce the received image via the medium of pencil on

[39] Schaffer, 'drawing', 457–459.
[40] Schaffer, 'Leviathan', 214, 216.
[41] Schaffer, 'drawing', 462.

paper in conditions of low illumination (in order to avoid ruining their adjustment to the limited light made available through the telescope).[42] And seeing itself, as noted by William Herschel, is an art which not only has to be cultivated but also diligently practised, suggesting that the brain's interpretive processes are at least as important as the image which was projected onto the retina.[43] South, for example, remained sceptical about some of the claims of nebular resolution made by his fellow observers:

> The night of the 5th of March, 1845, was the finest I ever saw in Ireland. Many nebulas were observed by Lord Rosse, Dr. Robinson and myself. Most of them were for the first time since their creation, seen by us as groups or clusters of stars; while some, at least to my eyes, showed no such resolution.[44]

Further transformations took place as the original illustrator subsequently conveyed to the engraver an interpretation of how the objects and textures in a drawing should be rendered on a printing plate. It was, in fact, only an accompanying narrative and the authority of its author that determined whether individual dots or textures on the page should be understood as clouds of stars or a field composed of a nebulous (i.e. gaseous) substance. As Nichol himself put it at this time:

> In reference to the Plates in this volume, I must acknowledge my great obligations to Lord Rosse. The remarkable Spiral Nebula is now published for the first time, through his kindness; and I am glad to state, that aided by willing and ingenious artists my rather venturous attempt to represent these masses of stars in the light in which they appear—viz: *white on a dark ground*, has been considered by his Lordship to be successful.[45]

It is worth pointing out at this stage that the Birr Castle observations were taking place at the beginning of the period during which, as Lorraine

[42] Omar W. Nasim, 'The 'Landmark' and 'Groundwork' of stars: John Herschel, photography and the drawing of nebulae', *Studies in History and Philosophy of Science*, 42 (2011): 67–84 at 70–74.

[43] Schaffer, 'Leviathan', 190–191.

[44] Excerpt from a letter written by Sir James South which was printed in the *Times* of 16 April 1845 and in the *Illustrated London News* of 19 April 1845. Reprinted in Thomas Dick, *The Practical Astronomer* (London: Seeley, Burnside and Seeley, 1845), 555.

[45] John P. Nichol, *Thoughts on Some Important Points Relating to the System of the World* (Edinburgh: William Tait, 1846), x.

Daston and Peter Galison have pointed out, a slow transition was in progress from what they have termed "true to nature" representations of the natural world towards the "epistemic virtue" of "mechanical objectivity". To achieve this desirable state, it was beginning to be felt necessary to suppress those elements of the self which might be tempted to impose personal hypotheses on observational or experimental data.[46] It would appear however, on the basis of Robinson's reported comments in Dublin, that he at least was more inclined to give priority to the subjective needs of perceived religious truth than to mechanical objectivity.

In the particular case of nebular astronomy, Omar Nasim has pointed out that the move towards mechanical objectivity was facilitated by changes in the technologies employed by observers and illustrators. He contrasted the loosely defined representations of nebulae produced by William Herschel with the ways in which his son developed new drawing and plotting techniques, supported by improved papers and pencils, to produce precise and detailed images of such objects as the Orion nebula:

> When one compares some of these exquisite graphite pencil drawings made by Herschel [using a *camera lucida*] with those pencil drawings he was to later make of the nebulae, one instantly recognizes a continued enthusiasm for minutiae; for an abundant, individual, and detailed depiction.[47]

Lord Rosse may well have conceived that he was pursuing his telescopic observations with a degree of objectivity and in accordance with the precedent set by Herschel. In practice, however, much of the work was subcontracted to paid assistants and his astronomical ledgers, whilst providing a voluminous longitudinal survey of many observed objects, did not for the most part yield diagrams which could rival, in terms of detail and precision, those of his English competitor.[48] He did nonetheless manage to produce one or two very successful 'broad brush' illustrations, especially one of the spiral nebula Messier 51 which, as noted above, was first

[46] Lorraine Daston and Peter Galison, *Objectivity* (New York: Zone Books, 2010), 17–19, 213–234.

[47] Omar W. Nasim, 'Observing by hand', *Rheinsprung 11-Zeitschrift für Bildkritik*, 3, (May 2012): 66–74 at 69.

[48] Omar W. Nasim, 'Extending the Gaze: The Temporality of Astronomical Paperwork', *Science in Context*, 26, Special Issue 2 (June 2013): 247–277 at 264–266. https://doi.org/10.1017/S0269889713000057.

published in 1846 by Nichol who saw it, in keeping with his cosmogony, as further evidence of stellar evolution:

> This plate, which is copied from the sketch shown by his lordship at the recent meeting of the British Association, (1845) is an eye sketch only unverified by micrometrical measurements, though these, when made, will alter no essential feature.
>
> the conversion of these dim streaks of light into burning and rolling orbs, has an attraction infinitely less than the transforming of a shape apparently simple, into one so strange and complex that there is nothing to which we can liken it, save a scroll gradually unwinding, or the evolutions of a gigantic shell![49]

Rather than relying on reproductions of nebulae, Rosse and Robinson were able to utilise two other key resources for managing truth claims. Ownership and control of the telescope determined what could be seen, and by whom it could be seen. Access to consumers of the written and spoken word in Ireland and Britain determined not only what could be reported but also how their reports were to be moulded to fit their intended audiences. The impassioned sermons delivered by Robinson to a partisan audience in Dublin differed markedly from the more dispassionate and objective tone of the comments made by Rosse to the London press and the addresses he gave in his role as president of the British Association for the Advancement of Science from 1843 onwards.[50] The production of epistemic virtue was conditioned by the combination of the scientific self with the "attitudes, values, and social relations that operated in specific locales" including the "London gentlemen of science" and the more heterogenous audience of the Royal Irish Academy.[51]

It is in any case clear that Robinson used his verbal descriptions of heavenly objects to consistently achieve results which conveniently accorded with his personal theological perspectives rather than making any serious attempt to capture mechanical representations. Or, to put it more vernacularly, he tended to see what he wanted to see:

[49] Nichol, *Thoughts*, Plate VI between pages 22 and 23, 23–24.

[50] Schaffer, 'Leviathan', 211. Parsons (ed.), *scientific*, contains inter alia reprints of five papers presented by Robinson to the Royal Irish Academy and seven papers presented by Rosse to the British Association for the Advancement of Science.

[51] Daston and Galison, *Objectivity*, 202.

> With earlier experiences of the bad Irish weather, the unusually perfect sky made him hurry up. Fully programmed for success, he wanted to get the desired results as soon as possible. Thus a maximum number of nebulae had to be 'resolved'.[52]

There may even have been at least one instance of deliberate 'sleight of hand'. Wolfgang Steinicke has asserted that Robinson's tacit intentions were supported not only by his confident predictions of favourable future results but also by subtle manipulation of the tools used in observation.[53] It is suggested not only that Robinson succeeded in convincing himself of the purely stellar nature of the spiral nebula Messier 51 but also that he then actively deceived his fellow observer South by suitable selection of eyepieces to convey the impression that it is a globular cluster of stars with no hint of nebulosity, whereas a wider field of view would have revealed the spiral arms and an attendant smaller cluster which would had given it, as noted above, a much greater appearance of dynamism and nebulosity.

Artistic Responses to the Leviathan and the Orion Nebula

It must have occurred at some point to Robinson that the superior light-gathering power of the Leviathan might also be capable of establishing whether the universe had limits and thus resolve an uncertainty which had occasionally also exercised Nichol: Is the cosmos infinite in extent and therefore by implication also in time? He and Rosse did not, however, discover any perceptible boundary to creation even at the telescope's highest magnification and at distances which were vast by preceding standards. Why was there no end to the universe and where, if anywhere, did God reside? It was not possible, even with the most powerful instrument thus far developed, "to behold the beauty of the Lord", although that very act, if feasible, would have been fraught with a danger which would also have been well known to students of the Old Testament.[54]

[52] Wolfgang Steinicke, 'Birr Castle observations of non-stellar objects and the development of nebular theories' in Mollan (ed.), *Parsons*, 210–270 at 231. Steinicke, 'The M51 mystery'.

[53] Steinicke, 'The M51 mystery' 19–29.

[54] Psalm 27:4. Exodus, 33:20. "And he said, Thou canst not see my face: for there shall no man see me, and live."

Although the suggestion that Robinson might have held out such an exotic expectation for the newly constructed telescope might seem a priori unlikely, we should bear in mind some of the elegiac verses with which he ushered in this neonatal aristocratic instrument, the result of the heroic "Intellect and Might" of his patron:

> Welcome to thy new existence,
> Child of Intellect and Might!
> Welcome to thy wide Dominions,
> Lord of Ether and of Light!
>
> Thou shalt lead us on in triumphs,
> Yet to mortal power unknown;
> Realms, which Angels only visit,
> Shall yield homage at thy throne.[55]

In contrast with Robinson's elegiac poetry, we have the imaginative and fearful description of the Orion nebula, which appears in the epigraph at the head of this chapter. Thomas De Quincey (1785–1859) mistakenly supposed he had based his comments on a drawing of the object as seen through the Leviathan. He was in fact referring to a detailed drawing produced earlier by John Herschel.[56] De Quincey had an established record as a writer of surreal essays which he himself declared to be based on the "dreams" of an "opium eater".[57] The effects of his opium consumption, a habit which Nichol in due course also acquired, would have contributed significantly to the sense of expanding cosmic time and slowness of interstellar motion which he had earlier encountered in his dreams.[58] But his conclusion that the nebula contains "a horrid chasm, a ravine, a shaft, that many centuries would not traverse" adds a somewhat darker colour to his description.[59] De Quincey's interest in the results of observations and comments made by astronomers, including his close friend Nichol,

[55] Bennett, *Church*, 107.

[56] Jonathan Smith, 'De Quincey's Revisions to "The System of the Heavens"', *Victorian Periodicals Review*, 26, 4 (Winter, 1993): 203–212.

[57] Thomas De Quincey, *Confessions of an English Opium-Eater*, Second Edition (London: Taylor and Hessey, 1823), 11–12.

[58] De Quincey, *Confessions*, Second Edition, 158–160. John P. Nichol, *Memorials from Ben Rhydding: Concerning the Place, Its People, Its Cures* (London: Charles Gilpin, 1852), 1–5.

[59] De Quincey, 'System', 571.

enabled him to sense that the universe is a vast, ancient and potentially hostile environment which no longer gave the impression of having been created solely or even primarily for the benefit of human beings.[60]

Astronomers had at that time only recently succeeded in measuring the parallax (and hence the distance) of two nearby stars.[61] The nearer of the two, α *Centauri*, had been found to lie at least three light years away from the solar system (a suggested measure of sidereal distances expressed in units of time popularised by Herschel in his *Treatise*), implying that all other stars must be even farther away from the earth and presumably similarly distant from each other.[62] It was evident to astronomers, as it had been to De Quincey, that the scale of the totemic Orion nebula, which contains a number of clearly resolved stars in its central region, is very much larger than any single planetary system. If, in fact, the apparently gaseous clouds in the nebula are formed of a veritable mass of stars which are individually almost too faint to be resolved by the world's most powerful telescope, then the nebula must be very distant indeed: a thought which was subsequently to be exploited by Nichol. Rosse and Robinson, on the other hand, seem to have, deliberately or otherwise, overlooked such vast variations in scale so that the resolution of Orion could also be used to support the refutation of the planetary evolution hypothesis employed by the author of *Vestiges*.

Nichol's Initial Response to the 'Resolvability' of the Orion Nebula

One rarely discussed feature of the Leviathan episode is the way in which both variants of the nebular hypothesis were deemed by a number of commentators to critically turn on a single point of evidence: the resolvability of the totemic Orion nebula.[63] It may be recalled that he primary supporting evidence for the sidereal nursery component of the nebular hypothesis

[60] Alex Murray, 'Vestiges of the Phoenix: De Quincey, Kant and the Heavens', *Victoriographies* 1.2 (2011): 243–260. https://doi.org/10.3366/vic.2011.0031.

[61] 61 *Cygni* and α *Centauri*—the measurements were made by Friedrich William Bessel (1784–1846) and Thomas James Henderson (1798–1844), respectively.

[62] Herschel, *Treatise*, 377–379, and John F. W. Herschel, *Outlines of Astronomy* (Philadelphia: Lea & Blanchard, 1849), 506–513. See also Thomas Dick, *The Sidereal Heaven and Other Subjects Connected with Astronomy* (New York: Harper & Brothers, 1840), 68–71.

[63] Schaffer, 'Leviathan', 199–200.

consisted of the observations of nebulae made by the Herschel family. They found that a significant number of the 2000 or so nebulae that they had catalogued appeared to consist of clouds of gas or 'shining fluid' Of all the objects surveyed, the Orion nebula is one of the largest, is relatively easy to find with the naked eye, and has a highly irregular appearance which had apparently exhibited "marked changes" and which distinguished it from more resolvable objects such as globular clusters as noted by Nasim:

> Due to its distinct irregularity in form and its unique appearance, this nebula had been suspected, at least since the time of William Herschel, of being the. key to the determination of whether or not a self-luminous nebulous fluid existed in the nebulae.[64]

Thus, the Orion nebula, like some of its peers, could potentially be seen as a nascent stellar system: an indication that the evolution of the universe was still in progress. Opponents of the concept that creation was incomplete, imperfect and capable of being modified by processes which had the appearance of random ebb and flow, like the Brownian motion of dust motes, were therefore keen to demonstrate that no such irresolvable nebulae existed and that they were all in fact collections of completely formed stars.[65]

In the context of the different political and theological viewpoints which were at that time entangled with the verifiability of the nebular hypothesis, it is perhaps surprising that the first person to have had an opportunity to view the Orion nebula through the six-foot reflector had been Nichol, who was a particularly enthusiastic advocate of developmental cosmogony. Nichol by now had an apparently amicable relationship with Lord Rosse, to the extent that he had previously courted controversy within the University of Glasgow by agreeing to loan him equipment originally intended for use in the university's own observatory.[66] The unintended consequences of this action included, in due course, significant

[64] Schaffer, 'Leviathan', 199. Omar W. Nasim, 'Observation, working images and procedure: The 'Great Spiral' in Lord Rosse's astronomical record books and beyond', *The British Journal for the History of Science*, 43, 3 (September 2010): 353–389 at 380. https://doi.org/10.1017/S0007087410000051.

[65] Iwan R. Morus, *When Physics became King* (University of Chicago Press, 2005), 202–208.

[66] James Coutts, *A History of the University of Glasgow: From its Foundation in 1451 to 1909* (Glasgow: James MacLehose and Sons, 1909), 388–389.

problems in his relationship with his employer and financial failure.[67] His 'generosity' seems particularly surprising when viewed in the light of his earlier quite trenchant views on the iniquitous behaviour of some members of the protestant ruling class in Ireland and on the use of tithes to support the established churches (which would have put him in opposition to Robinson, whose post at Armagh was partially funded by such tithes).[68]

Nichol was clearly also at odds with Robinson's requirement for a religiously founded 'static' cosmology. That Robinson was likewise strongly opposed to Nichol's "zeal for Nebular Cosmogony" is clear from an 1841 letter to Rosse, who "often had to serve as a mediator" between the two rivals. Robinson's flattering comparison, in which he imagined Nichol being "in despair of what he hears you have done", also provided a pointer to the practical difficulties which duly arose in the Glasgow observatory:

> Nichol, professor of Astronomy at Glasgow University, the author of a catchpenny book which you may have seen called "The Architecture of the Heavens". He seems in despair of what he hears you have done, for in his zeal for Nebular Cosmogony, he had acquired a mirror by Ramage of 2 feet aperture and 50 feet focus! with which he has to do wonders.[69]

Over time Nichol became aware of the unsuitability of the Ramage mirror "with which he has to do wonders", and he was therefore keen to better understand the impressive new technology about to be deployed in Ireland.[70] His donation of astronomical apparatus helpfully altered his status vis-à-vis the Irish project to that of sponsor, thus facilitating access to the newly completed telescope. Nichol was also motivated by a conviction that the observations to be made by the Leviathan would provide confirmatory evidence in favour of his preferred cosmogony. And, as we have seen, Nichol was quite adept at adapting to the requirements of the various contrasting, and sometimes conflicting, environments in which he found himself.

[67] Schaffer, 'nebular', 141.

[68] Bennett, *Church*, 80, 87.

[69] Robinson to Rosse, 7 April 1841, Birr Castle MSS K 5.4. Steinicke, *Observing*, 105.

[70] His son's account of the difficulties encountered in using the 50-foot telescope and its subsequent destruction in a storm can be found in: William A. Knight, *Memoir of John Nichol, Professor of English Literature in the University of Glasgow* (Glasgow: James Maclehose, 1896), 33–35.

The ambiguous nature of the relationship between Nichol and Rosse also illustrates the complexity of the interaction between academic and aristocratic authority in the mid-nineteenth century, especially given the traditional overlap between these two groups. Rosse appeared to accept at least a partial equivalence in Nichol's structural status following his donations, permitting a level of rapprochement between the astronomical rivals. Nichol was in due course invited to Birr Castle and offered the opportunity to view the Orion nebula in December 1845. Following his night of observation, as we saw in Chap. 6, Nichol declared that he was unable to resolve the nebula into individual stars.[71]

On receiving a subsequent report, in a letter from Rosse, of its probable resolvability, which he interpreted as actual resolution, Nichol publicly accepted the word of a gentleman with respect to this new evidence: an acceptance which reflected the practical impossibility of challenging the veracity of a peer of the realm.[72] A second factor which may have played a role in Nichol's relatively ready acquiescence to the Rosse interpretation of the observations of Orion was another unintended outcome of his significant but problematic material support for the project. Having by then returned to Glasgow, he probably did not want to engage in further discussion of the extent or utility of his ill-advised equipment 'loans'.[73] As we shall see in Chap. 8, however, Nichol refused to abandon the idea of nebular cosmogony, and, in fact, he was subsequently to reintroduce it as evidence for the great age and extent of a universe which he was sure was, is and would continue to be in a state of evolution at all scales.

[71] John P. Nichol, *The Architecture of the Heavens* (London: John Parker, 1850), 113–114.

[72] Schaffer, 'nebular', pp. 141–142. Also see James A. Secord, *Victorian Sensation: The Extraordinary Publication, Reception, and Secret Authorship of Vestiges of the Natural History of Creation* (University of Chicago Press, 2000), 403–410, for a discussion of the role of gentlemen in science.

[73] Coutts, *History*, 388–389.

CHAPTER 8

"In Tracts of Fluent Heat Began": Nichol Defends and Develops Nebular Cosmogony

They say,
The solid earth whereon we tread

In tracts of fluent heat began,
And grew to seeming-random forms,
The seeming prey of cyclic storms,

Alfred Lord Tennyson, 'In Memoriam A. H. H'., CXVIII'[1]

In his pursuit of a progressive cosmology in which stellar systems had evolved from clouds of gas rather than having been set in place at the beginning of time by a creator, the reported resolvability of the Orion nebula clearly was, at least initially, a setback for John Pringle Nichol (1804–1859) and other supporters of the nebular hypothesis. One contemporary commentator observed, "Without doubt then, the nebular hypothesis must be abandoned", relying in part on the news that Nichol had himself written to the editor of the *Glasgow Argus* to report that the concept "is no longer tenable".[2] Despite his apparent concurrence with the Irish observers, however, Nichol was unwilling to give up the idea of a developing cosmos. He employed his ability to connect with the audiences

[1] Alfred Lord Tennyson, 'In Memoriam A. H. H.' in Karen Hodder (ed.), *The Works of Alfred Lord Tennyson* (Ware: Wordsworth Editions, 1994), 305–389 at 379.

[2] 'Journal of Science, Inventions, and improvements', *The Critic*, 1844–1847; Apr 11, 1846; 3, 67: 405.

H. Carlton, *Cosmology and the Scientific Self in the Nineteenth Century*, https://doi.org/10.1007/978-3-031-05280-4_8

for his popularisations of current astronomical thinking to promote new variants of his previously stated positions.

In 1846, shortly after the reported resolution, he published *Thoughts on Some Important Points Relating to the System of the World*: a major statement of his continuing belief in evolutionary hypotheses. The book began with apparently fulsome praise for the achievements of Lord Rosse together with a concession that the sidereal nursery component of nebular cosmogony might now be a 'dead letter':

> In the first place, the action, although only begun, of the great telescopes which Science owes to the genius and labour of the Earl of Rosse have somewhat altered the views I formerly gave to the public [...] witness the extraordinary revelations regarding the shapes and internal constitution of the Stellar clusters, which, through the kindness of that Noble Earl, I am enabled very fully to present: but, in regard of one special and very important point, his Lordship has wholly subverted the opinion of his illustrious predecessor [William Herschel].[3]

This approach is a further example of a literary device which he had previously employed in several debates, including his disagreements with Thomas Chalmers (1780–1847), that of hyperbolic flattery followed by subtler demolition. In this instance, following the fulsome introduction, he subsequently proceeded to undermine the inference drawn by Thomas Romney Robinson (1792–1882) and Lord Rosse (1800–1867) that the nebular hypothesis had been totally disproved.

As Nichol's arguments unfolded through the course of *Thoughts*, he gradually challenged the solidity of the Irish observations and the conclusions drawn therefrom by means of contrary evidence such as the 'planetary' nebulae discovered by Frederick William Herschel (1738–1822), the persistent unresolvability of the Andromeda nebula, the mysterious zodiacal light and, more generally, the continued applicability of Laplace's separate idea of planetary formation. By the end of Chapter II (and its accompanying Appendix), the reader had been informed that the idea of a nebular origin for the solar system was still applicable and was well prepared for the subsequent rehash of some of the geological and biological

[3] John P. Nichol, *Thoughts on Some Important Points Relating to the System of the World* (Edinburgh: William Tait, 1846), v–vii.

developmental ideas which had also featured prominently in *Vestiges of the Natural History of Creation*.[4]

The reviewer of *Thoughts* in *The Spectator* concurred that "La Place's theory of the formation of our solar system" still stood, that one could only speak of an "alleged dissolution" of the nebular hypothesis and rather generously credited Nichol's defence of his own ideas as an attempt to exonerate Herschel senior "from any suspicion of rashness in forming his speculation"[5]

Nichol Reframes the Birr Castle Observations

But the key argument which Nichol deployed at this point, and subsequently repeated in a series of public lectures delivered at such geographically dispersed locations as the Athenaeum club in London, the Manchester Royal Institution and the Mercantile Library Association of New York, was a rather ingenious reframing of the Orion nebula observations to support his views regarding the scale and age of the universe.[6] To summarise his argument as succinctly as possible, Nichol held that if the previously unresolved nebulae consist ultimately of clusters of stars which are only just resolved by the Leviathan, then the distance of these stars from earth is considerably greater than previously supposed to the extent that the Orion nebula must be at least 60,000 light years away and correspondingly massive in scale. He asserted that the furthest visible objects contained in Orion and elsewhere in the sky are clusters which appear to us as points of light. They are therefore perhaps 30 million light years away (and ago), as noted by the reporter from the *Manchester Guardian*: "About 5000 of these clusters had been observed, from some of which the light occupied 30 million years in coming to us."[7]

[4] Nichol, *Thoughts*, 39–47, 56–86.

[5] 'Dr. Nichol on the system of the world (Book Review)', *The Spectator*, 12 September 1846; 19, 950: 882–883 at 882.

[6] The lecture series at the Athenaeum was reported by *The Manchester Guardian* editions of 12, 15, 22 and 26 May 1847. The lectures in Manchester were similarly covered by the same journal in its 10 and 24 September 1851 editions. The series delivered in New York was published in book form as: John P. Nichol, *Views of Astronomy: Seven Lectures Delivered Before the Mercantile Library Association of New York in the Months of January and February 1848* (New York: Greeley & McElrath, 1848).

[7] 'Professor Nichol's lectures on recent astronomical discoveries' *The Manchester Guardian*, 12 May 1847.

In making this assertion, Nichol was also, after Herschel senior, one of the first people to suggest that our own galaxy was just one, perhaps typical, representative of the clusters of stars which are repeated right across the cosmos: a significant expansion of the scale and age of the universe, implying a negative step change for the relative importance of the inhabitants of any individual planet. Not only is the earth not the centre of the solar system, nor is the sun a particularly special star but, according to Nichol, the Milky Way may be just one of many such "superb groups or galaxies":

> separated from each other by gulfs so awful, that they surpass the distances which divide star from star, even as far as these transcend the spans which, as our terrestrial units, enable us to survey the magnitudes that concern us on the surface of our own abode.[8]

Nichol had executed a classic Morton's fork: if the Orion nebula is unresolvable, then Herschel's developmental hypothesis had been confirmed but, if not, the universe is far older and more extensive than anyone had previously thought. In either case, the planetary speculation of Pierre-Simon Laplace (1749–1827) was not in his view, and that of some reviewers, disproved by the findings of the Irish astronomers.[9]

Nichol's novel suggestion for how the possible resolutions of nebulae were to be interpreted allowed him to reclaim his astronomical authority in a series of public pronouncements. He was now able to restate his prior positions on cosmic development and also to significantly enlarge the known universe. This rescaling of the cosmos was, moreover, broadly consistent with the apparent ages of the rocks in the earth's crust. Thus, Nichol neatly contrived to turn the evidence which had initially seemed to destroy the nebular hypothesis into yet another leg of support for theories

[8] John P. Nichol, *The Stellar Universe: Views of its Arrangements, Motions, and Evolutions* (Edinburgh: John Johnstone, 1848), 59.

[9] A Morton's fork is a specious piece of reasoning in which contradictory arguments lead to the same (unpleasant) conclusion. It is said to originate with the collecting of taxes by John Morton, Archbishop of Canterbury in the late fifteenth century, who held that a man living modestly must be saving money and could therefore afford to pay taxes, whereas if he was living extravagantly, then he was obviously rich and could definitely afford them. Hannes Kleineke, "'Morton's Fork'? – Henry VII's 'Forced Loan' of 1496'", *The Ricardian*, xiii (2003): 315–327.

of cosmogony and geology which also pointed to ideas of change and progression in the spheres of biology and society.

That this was indeed Nichol's intention can be seen in his expositions of the duration and complexity of both terrestrial and extraterrestrial geological developments in *Thoughts* and his subsequent lecture series. The implied connections between distance and time and the links to developmental hypotheses were also clearly perceptible to his audiences. As *The Spectator* commented:

> The enormous duration of the system, measured by our notions of time—the state of continual change which is everywhere traceable [...] are the main topics. In the course of their discussion, Dr. Nichol brings together some of the more striking facts both of astronomy and geology, making them his own by applying them to his exposition of Earth's cosmogony: but in looking at the vastness and duration of the material world, he does not forget the dignity of the moral.[10]

Or as George Gilfillan (1813–1878), with a characteristic injection of florid description and anthropomorphic theology, put it:

> for now we have been taught, in a manner most impressive, the immense age of the universe, whose orbs seem hoary in their splendour, and have thus formed a new measure for computing our knowledge, or rather for more accurately estimating our ignorance of the days, of the years, of the right hand of Him that is the Most High.[11]

It is evident from both reviews that Nichol continued to couch his arguments in a context which provided a linkage to the sort of designing and creating deity posited by natural theologians, without being overly specific about the nature or role of such a divinity. A more overtly materialist treatment of astronomy and geology would have been too dislocating for some audience members and could have attracted censure from those "saints" whom Robert Chambers (1802–1871), the author of *Vestiges*, had felt it necessary to assuage.[12]

[10] 'Dr. Nichol', *The Spectator*, 883.

[11] George Gilfillan, *A Second Gallery of Literary Portraits* (Edinburgh: James Hogg, 1850), 235. In this passage, Gilfillan was paraphrasing Psalms 77:10.

[12] See Chap. 6.

Nichol's Lecture Tours

Nichol had become an accomplished actor in the virtual pulpit of public performance and could be compared favourably with his predecessor Chalmers, a noted orator in his own right. James MacLehose (1811–1885), a Glasgow publisher, observed in a brief pen portrait of Nichol that although separated in time by a quarter of a century, or roughly one generation, both men were known for their ability to command the attention of large audiences even when, in Nichol's case, discussing "minute processes" rather than leading them "into realms of illimitable space":

> when he spoke rather of minute processes than general results the attention was as close and the interest almost as keen as when he led his listeners into realms of illimitable space. Since the day when Chalmers delivered his sermons in the Tron Church no public speaker north of the Tweed had so enchained his hearers. One who was himself a most accomplished lecturer was wont to say that all that the people of Scotland knew of astronomy they had learnt from Thomas Chalmers and John Pringle Nichol.[13]

Nichol's motivation for conducting his extensive lecture series seems to have not only been an intellectual crusade designed to persuade a wider audience of the validity of his views on the nature of the universe. He was also actuated by the need to recoup the financial losses brought about by the overliberal investment in the University of Glasgow observatory (and thence into Rosse's telescope) which had "sent him into irretrievable debt".[14] His pursuit of both of these objectives was greatly assisted by his ability as a speaker "with the power of an accomplished orator". Nichol's "lectures were works of art, calculated at once to exercise and enlighten the intellect, quicken the imagination, and move the heart": aptitudes which he had exhibited from an early age.[15]

Although his subject matter gave the external appearance of being grounded in objective science, he shared with both Chalmers and contemporary exponents of popular astronomy such as Ormsby MacKnight

[13] James MacLehose, *Memoirs and Portraits of One Hundred Glasgow Men* (Glasgow: James MacLehose & Sons, 1886), 250.

[14] Simon Schaffer, 'The nebular hypothesis and the science of progress' in J. R. Moore (ed.), *History, Humanity and Evolution: Essays for John C. Greene* (Cambridge University Press, 1989), 131–164 at 150.

[15] MacLehose, *Memoirs*, 250.

Mitchel (1810?–1860) the ability to freely access a performative self. Both Nichol and Mitchel adopted a pseudo-religious delivery style, verging on the elegiac, including references to the powerful attributes of the deity at key points, particularly in their introductory and concluding statements.[16] Nichol's lectures were delivered in tones of reverential awe which provided a continued linkage to the vague but satisfying concept of a creative genius, an "Infinite Essence" which he conceived of as being the orchestrator of the universe:

> No! above, below, around—there is GOD; there, his universal presence, speaking to finite creatures in finite forms a language which only the living heart can understand. [...]—there is the Infinite Essence, and the mystic development of His Will![17]

An additional reading of Nichol's approach is that he permitted himself, as a result of his breakdown, to elide his personal theology with the external trappings of astronomy, thus allowing him to deliver a series of public sermons on sublime and celestial themes which were consistent with his somewhat unconventional *Weltmodell*.[18]

A surprising attribute of Nichol's lecturing style was that he suffered from a slight speech impediment, probably a stammer. Reactions to such an apparent speech defect in the nineteenth century were not as negative as we might suppose. According to Josephine Hoegaerts, it was believed that stammerers "were typically found in the professional classes and characterized by an extraordinary intelligence".[19]

In Nichol's case, his "hesitation" was apparently an endearing attribute which enhanced his performances. It was positively received by Gilfillan,

[16] Frederick A. Mitchel, *Ormsby MacKnight Mitchel: Astronomer and General. A Biographical Narrative* (Boston: Houghton, Mifflin, 1887), 161–162. Ormsby M. Mitchel, *Popular Astronomy* (London: George Routledge and Sons, 1860), 290. Examples of Nichol's theological interjections can be found in John P. Nichol, *The Stellar Universe: Views of Its Arrangements, Motions, and Evolutions* (Edinburgh: John Johnstone, 1848), vii–viii, 223–227.

[17] Nichol, *The Architecture of the Heavens* (London: John W. Parker, 1850), 219.

[18] A German word which has previously been employed to describe both the cosmological models of physical scientists and the creation stories found in traditional religions. It thus encompasses the two key epistemologies under discussion here.

[19] Josephine Hoegaerts, 'S-s-s-syncopation: Music, Modernity, and the Performance of Stammering (Ca. 1860–1930)', *Societies*, 2015, 5: 744–759 at 750. https://doi.org/10.3390/soc5040744.

who interpreted "a certain hesitation" when he engaged in "extempore speech" as a sign of the unusual productivity of his mind in much the way that Hoegaerts suggests:

> His manner, without detracting aught from, adds little or nothing to, the impression of his thought or style, of which it is simply the medium. Its principal quality is ease—an ease not materially impaired by a certain hesitation. Hesitation, we need scarcely say, has often a great charm. […] And we know some popular divines who have stammered themselves into pulpit celebrity […] Dr. Nichol's hesitation is not great, is confined to his extempore speech, and seems rather to spring from an excess than a deficiency of matter or words.[20]

Gilfillan, in the course of this effusive eulogy, also extolled Nichol's ability to "uncase science from its mummy confinements, and to make it walk abroad as a free and living thing". This striking analogy would have been understood by his readership as being derived from the contemporary vogue for Egyptomania. It referred to the "intoxicating experience" of "mummy unwrapping spectacles [which] allowed bandages and trinkets from the body to be inspected, touched and smelled".[21]

Nichol's son John (1833–1894) confirmed that it was the combination of his father's imaginative language and the "power" of his speech which were the keys to his performance and which he contrasted favourably with his own more modest efforts:

> I was familiar with most of those facts and theories concerning the earth, and moon, and stars, which a large audience came to learn, for twenty successive weeks, in the Merchants' Hall. Without that power which could at will arrest and make wonder the eyes of his hearers; without that grace which gave to his speaking the charm of a perfect actor; without that fire of eloquence which raised me into another world every time I heard it, I could have delivered most, as I was once constrained to deliver one, of my father's lectures to a disappointed audience.[22]

[20] Gilfillan, *Second Gallery*, 254.

[21] Mummy-unwrapping parties took place from the early nineteenth century onwards but surged in popularity in the 1830s when a number of such events were hosted by a London surgeon called Thomas Joseph "Mummy" Pettigrew (1791–1865). See Eddie Tapp, 'The Unwrapping of a Mummy' in A. Rosalie David (ed.), *Manchester Museum Mummy Project: Multidisciplinary Research on Ancient Egyptian Mummified Remains* (Manchester University Press, 1979), 83–94. Eleanor Dobson, 'Literature and Culture in the Golden Age of Egyptology', PhD Thesis submitted to the University of Birmingham (2016), 31.

[22] William A. Knight, *Memoir of John Nichol, Professor of English Literature in the University of Glasgow* (Glasgow: James Maclehose, 1896), 70.

Reception of Nichol's Presentations

Two interesting points emerge from Nichol junior's comments. Firstly, we gain an impression of how popular, and presumably lucrative, his father's lectures were, attracting "a large audience" which "came to learn, for twenty successive weeks". Secondly, and even allowing for a little exaggeration on his son's part, it is also evident that his talks were as much appreciated for their affect, "that fire of eloquence which raised me into another world every time I heard it", as for his "facts and theories".

It was also a characteristic of Nichol's public lectures that he deployed large (and more conceptual than detailed) illustrations to accompany his sweeping statements and grandiloquent claims: audience-pleasing techniques subsequently adopted by Richard Anthony Proctor (1837–1888).[23] His effusive style and hyperbolic statements, however, also drew some criticism from his peers as Simon Schaffer has pointed out:

> Several critics found Nichol's 'gorgeous style, gigantic diagrams and enthusiasm' rather suspect, both politically and morally. His 'very liberal and very unconsidered use of such words as inconceivable, infinite, eternal' simply deprives them of all such power, and at last for the infinite we mentally substitute greater than earth and for eternal, longer than the earthly lifetime of man'.[24]

Nichol was, as noted above, not the only astronomical lecturer in circulation at that time nor was he atypical of the "commercial but sublime" segment of the market for popular science. In addition to a few established professionals "there were far more private showmen who engaged in the business of popular astronomy lecturing" earning a satisfactory stipend with their itinerant, theatrical performances, similarly accompanied by "beautiful visual representations" and purpose-built demonstration equipment. The sublime component of their output consisted, as did Nichol's, of interspersed allusions to a vague and naturalistic theology, references to the awe-inspiring scale of the cosmos and the impressive capabilities of the creator which stimulated a sense of affective awe.

The audiences for popular astronomy were composed of people of varying ages, class backgrounds, religious beliefs and political inclinations

[23] See Chap. 5.

[24] Schaffer, 'nebular', 150. The embedded quotations are from Gilfillan, *Second Gallery*, 255 and 'Recent Astronomy', *British Quarterly Review*, (6 Aug. 1847): 1–40 at 26.

who consumed astronomical lectures for reasons which were many and varied, including moral instruction, religious inspiration, pursuit of knowledge and entertainment.[25] Clearly, the heterogeneous mixture of people who attended the lectures of Nichol and other popularisers would not have had a singular response to the material presented, and their impressions of these productions would depend to a great extent on cultural context and personal subjectivities. Their reactions did not necessarily correspond to essentialist narratives of Victorian mentalities and the now-outdated discourse of a long-and-steady decline in nineteenth-century religiosity.[26] As the previously quoted reports from the *Manchester Guardian* have demonstrated, reviewers reported Nichol's words and conclusions with little or no evidence of dissent either in their authorial voice or in terms of audience reaction. The fact that four of the lectures in question took place at the Athenaeum is, ipso facto, evidence of their acceptability in more conservative circles, presumably assisted by his occasional genuflections to a generic deity.

Another evident outcome of the oratorical capabilities evinced by Nichol was his ability to convince less scientifically aware audiences to a more or less uncritical acceptance of his arguments and evidence, however tentative the latter may have been in practice. Thus, for example, Edgar Allan Poe (1809–1849) was able to interpret the work of "Dr. Nichols" as having both established the truth of his [Nichol's] own ideas and to have validated the earlier imaginings of Epicurus: "the vague guesses of that noble Greek had met with confirmation in the late nebular cosmogony."[27]

One question which must arise from a consideration of the history of the nebular hypothesis and the interpretation presented by Nichol to his publics on the resolvability of Orion and its consequences for stellar distances is why there was so little overt challenge to his restatement of the problem in his own terms. A probable factor is that early Victorian audiences had begun to segment themselves more clearly than had their predecessors, and it is therefore not realistic to assume the existence of a

[25] Hsiang-Fu Huang, 'Commercial and Sublime: Popular Astronomy Lectures in Nineteenth Century Britain', PhD Thesis submitted to University College, London, (2015), 17–21, 275–276.

[26] For an extensive reinterpretation of earlier hypotheses of nineteenth-century secularisation, see: Callum G. Brown, *The Death of Christian Britain: Understanding Secularisation, 1800–2000* (London: Routledge, 2001).

[27] Edgar A. Poe, 'The Murders in the Rue Morgue', *Tales* (London: Wiley & Putnam, 1846), 116–150 at 122–123.

homogenous population who interpreted astronomical information and the debates it engendered via a uniform 'filter' or a single communication channel. It is evident that reception of these concepts was influenced by the audience members' belief systems and their propensity to accept or reject positivist arguments which contradicted traditional theologies. Readers were exposed to different views of the observations made at Birr depending on which periodical they favoured, with their patronage ranging from the more conservative organs such as *The Times* and *The Athenaeum* to *Chambers' Edinburgh Journal* and other more progressive publications. Perhaps the most surprising reaction was that of the Catholic *Dublin Review* which "expressed the same careful scepticism towards the news from Parsonstown". It also compared the images of two nebulae produced by Rosse with earlier pictures of the same objects published by John Frederick William Herschel (1792–1871) and noted that "the observations hitherto made upon nebulae are too imperfect to form a safe foundation for any hypothesis".[28]

John Herschel: Development of a Recognised Authority

Despite the apparent confidence which had accompanied the initial reports of nebular resolution by Rosse and Robinson, doubts about the validity of their observations surfaced quite soon thereafter and they both turned to established authorities in pursuit of confirmation or rebuttal. A key establishment figure to whom they appealed for support and assistance was John Herschel, the son of the originator of the stellar nursery form of the nebular hypothesis, who had by the 1840s become an adviser to the government and a widely read author of books on matters scientific in general and of astronomy in particular. His reputation was in part based on the significant expansion of the "star gauges" of the heavens created by his father and his aunt Caroline Lucretia Herschel (1750–1848) which was informed by a lengthy expedition to South Africa in the 1830s during which he surveyed the skies of the southern hemisphere.[29]

[28] Simon Schaffer, 'The Leviathan of Parsonstown: Literary technology and scientific representation', in Timothy Lenoir (ed.), *Inscribing Science: Scientific Texts and the Materiality of Communication* (Stanford: University Press, 1998), 182–222 at 215–217.

[29] Günther Buttmann (tr. Bernard E. J. Pagel), *The Shadow of the Telescope: A Biography of John Herschel* (Guildford: Lutterworth, 1974), 92–111, 120–129.

Herschel had an unusually wide range of interests and a creative but complex *Weltmodell.* He was academically very successful, despite an early setback when the young Herschel, described by a visitor as a "delicate, blue-eyed boy", was bullied at Eton. "As Eton was close to Slough [the Herschel family home], his mother visited him often; upon becoming aware of the treatment he had to endure there, she had him removed from the school and educated at home until he entered Cambridge in 1809." The Irish astronomer Agnes Mary Clerke (1842–1907) reported that he graduated from St John's College in 1813 "as senior wrangler and first Smith's prizeman, and was immediately elected to a fellowship in his college. He was at this time described by the poet [Thomas] Campbell as 'a prodigy in science, and fond of poetry, but very unassuming'". With fellow undergraduates George Peacock (1791–1858) and Charles Babbage (1791–1871) he formed the Analytical Society with the intention of revitalising the study of mathematics by the adoption of Leibnizian calculus. "To these works [the output of the Society] was mainly due the restoration of mathematical science in England by introducing the differential notation and continental methods of analysis."[30]

Following his considerable success in academia, Herschel then experienced a period of considerable stress, a "personal crisis centered around problems of career [...] and conflicts with his father" during which his future seemed very uncertain.[31] He rejected his father's pragmatic plan for a financially secure position in the Anglican Church. Writing to a university friend, the future Anglican cleric John William Whittaker (1791–1854), not only was he dismissive of those who took up the profession for primarily pecuniary reasons but also quite sceptical about the malign influence of religious enthusiasts and the established churches on natural religion:

> For verily, the church is the Lord's beehive and many are the drones which infest the same [...] where did you ever see a religionist whose only object was to climb up into Heaven. _ He will never be contented without a tug at the rest, to drag them neck & heels after him, or clamber on their shoulders to earthly preferment. Join with me, Whittaker in the wish & prayer, that

[30] Silvan S. Schweber, 'John Herschel and Charles Darwin: A Study in Parallel Lives', *Journal of the History of Biology*, 22, 1 (Spring, 1989): 1–71 at 7–14. Agnes M. Clerke, 'Herschel, John William Frederick', *Dictionary of National Biography* (London: Smith, Elder & Co, 1885–1900), vol xxvi, 263–268 at 263.

[31] Schweber, 'Herschel', 9.

religion, as established by law may never entirely usurp the superiority & controul over religion, established by nature.[32]

In his correspondence during this period, he also described periods of "depression and despondency" interspersed with bouts of frenetic activity, dabbling in astronomy, chemistry, physics, literature and even a brief flirtation with a legal career.[33] His mood swings were accompanied by the regular, indeed daily, consumption of alcohol: "Dearly beloved Whittaker, I am gloriously drunk, d__n me, and hope you are the same."[34] His drinking can also be read as a way of coping with periodic bouts of painful shyness, reticence and self-reproach. A later friend, Maria Edgeworth (1768–1849), observed that he suffered from

> mimosa sensitiveness that shrinks from every touch not merely of blame, but even from the very intimacy he most wishes to have [...] In his earlier life when disappointed in love he shut himself up in darkness for I don't know how many weeks and would let no mortal see him and even at later times when vexed in friendship or when scientific things go wrong he betakes himself to darkness and solitude and in abstraction shuts himself up from the external universe.[35]

In 1816, when his father became seriously ill, he reluctantly realised that it was his responsibility to take up the 'family business' of astronomy. It is clear that an important factor in this decision lay in his response to the resignation and stoic acceptance of inevitable mortality which he observed in his father's declining condition:

> I cannot be insensible to the great alteration which has taken place in him. Indeed he seems to be fully aware of it himself—he speaks of it with the tranquil calmness of Horace's philosopher—exacto contendu tempore vitae, and I am convinced from many of his expressions that the moment is seldom absent from his thoughts which will terminate a long and I may say with

[32] Herschel to John William Whittaker, 2 July 1813, Library of St John's College Cambridge.

[33] Schweber, 'Herschel', 8–12.

[34] Herschel to John William Whittaker, August 1812 and 22 May 1813, Library of St John's College Cambridge.

[35] Maria Edgeworth to Harriet Butler, 1843, quoted in Schweber, 'Herschel', 13–14.

> some pride, an honorable existence [...] I wished when I sate down to have written you an amusing letter, but at my outset unluckily I took the downward path_ My own spirits are depressed even to extinction.[36]

Whilst it might be tempting to see in the resolution of this crisis a parallel with Nichol's earlier conversion from religion to astronomy, it is evident that Herschel's decision had a lot to do with mending his previously volatile relationship with his father. "He was led to astronomy by filial piety, in opposition to a spontaneous preference for chemistry and optics."[37] Whereas Nichol under stress had reverted to a form of religion which had been received prior to his education in the tenets of Presbyterian orthodoxy, Herschel had no clear theology to fall back on. His father's views have been described as "wholly contemplative" and Herschel's "own early religious ideas were also along partly pantheistic and partly deistic lines". Over time, and especially after his marriage to the daughter of a Scottish Presbyterian, Margaret Brodie Stewart (1810–1884), the younger Herschel gradually developed a weak form of theism which extended to "an assurance, that his existence will not terminate with the dissolution of his corporeal frame" and that he would ascend to "a future state of being".[38]

But this was by no means a complete conversion to orthodox religion. In 1865, whilst clearly distancing himself from what he saw as the "Pantheist" concept of an uncreated world, Herschel described to an ecclesiastical correspondent a panentheistic universe interpenetrated by a rather impersonal omnipresent "being" very similar to that which had been proposed by Nichol in 1846. The mechanical paradigm had become so dominant in metropolitan circles that Herschel felt the need to seek theological 'absolution' from the Bishop of Kingston for his affirmation of the spiritual origins of force in two articles which he complained had been censored by the editor of the *Fortnightly Review*. His omnipresent and omniscient "Deity" was a

[36] Herschel to John William Whittaker, 2 September 1816, Library of St John's College Cambridge. The Latin tag is actually a misquotation from Horace, *Satire* 1.1, 117. The relevant phrase can be loosely translated as: "content with his life, he can retire from the world".

[37] Clerke, 'Herschel', 267.

[38] Buttmann, *Telescope*, 172–173. John F. W. Herschel, *A Preliminary Discourse on the Study of* Natural Philosophy (London: Longman, Rees, Orme, Brown, Green & Longman, 1832), 7.

> being who is present in every point of space & at every instant of time;— who is cognizant of every movement & action of every particle of matter; who either once and for all originally endowed or at every instant continues to endow these particles with all the powers and properties they have.[39]

Like Nichol he had, in effect, made God somehow less relevant to, or having less immediate impact on, the work of astronomers by imagining the deity as all-pervasive.

John Herschel and the Nebular Hypothesis

Herschel's attitude to the nebular hypothesis began with a slightly reluctant acceptance of his father's idea of stellar evolution.[40] His correspondence on the topic of the Laplacian variant, however, suggests that Herschel's reception of his peers' contributions was partly shaped by the political and religious affiliations of his correspondents, but was also heavily influenced by his opinion of the logical consistency of their arguments. He took a strong dislike to *Vestiges*. He saw it as metaphysically controversial in that its thesis of a continuum of development from Laplace's planetary formation hypothesis through to the process of speciation was a 'bridge too far' even for Herschel's general acceptance of divine operation through secondary laws. Herschel was unwilling to accept any suggestion that the production of current state of the universe could have been the result of the operation of stochastic processes whose existence had not been empirically confirmed.[41] But he also disliked the lack of quality and consistency in its argumentation, declaring that "I think it one of the shallowest productions I have seen [...] and full of misstatements, and errors of construction and logic".[42]

His correspondent on this occasion was Adam Sedgwick (1785–1873), who also held a nuanced view of the nebular hypothesis: "I delight in the nebular hypothesis. It was a glorious speculation of your immortal Father's". Sedgwick likewise condemned *Vestiges* as "rank materialism—foul and fetid".[43] He subsequently went into print with his own criticisms

[39] Herschel to the Bishop of Kingston, 14 July 1865, Royal Society Herschel MSS HS 24.118.

[40] Buttmann, *Telescope*, 103.

[41] Schweber, 'Herschel', 31–32.

[42] Herschel to Sedgwick, 15 April 1845, Royal Society Herschel MSS HS 15.427.

[43] Sedgwick to Herschel, 11 April 1845, Royal Society Herschel MSS HS 15.425.

of *Vestiges*, buoyed by several arguments which he had first sounded out with Herschel in this exchange of letters. Intriguingly, in his discussion of errors in the hypothesis as expounded by Chambers, Herschel can think of ideas which would "go some way to help us out of the scrape [...] a legitimate working out of my Father's views of nebular aggregation. But it is not what is commonly called 'the nebulous hypothesis'".[44] Sadly, but perhaps predictably in the light of his tendency to diffidence in such matters, Herschel did not deign to publish his version of this component of the hypothesis which could also have provided further confirmation that his approach to this argument was founded primarily on his interpretation of physical laws rather than any parallel epistemology.[45]

Herschel also discussed this topic with John Stuart Mill, a progressive thinker and an agnostic with a distinctly negative attitude towards the established church who had published many articles by Nichol and other radicals.[46] The supposed proof of the nebular hypothesis provided by Auguste Comte (1798–1857) was rejected by Herschel on the basis of the circular logic on which it depended. Herschel was fairly scathing in his comments on Comte's argument: "a vicious circle in the reasoning or some equivalent fallacy".[47] That he was unhappy with the illogicality of Comte's argument was also made clear in the speech to the 1845 British Association meeting at Cambridge which had originally prompted the writing of Mill's letter to him:

> where, I would ask, is there a student to be found who has graduated as a Senior Optime in this University, who will not at once lay his finger on the fallacy of such an argument, and declare it a vicious circle.[48]

[44] Herschel to Sedgwick, 5 April 1848, Royal Society Herschel MSS HS 15.438. Presumably, Herschel here is again making a distinction between his father's hypothesis and Laplacian planetary cosmogony.

[45] Herschel's eclectic, sometimes diffident approach to matters scientific is discussed in both Buttmann, *Telescope*, 55–57, 117–118, and Schweber, 'Herschel', 13, 68–69.

[46] William Stafford, *John Stuart Mill* (Basingstoke: Macmillan, 1998), 64–65. An example of an article written by Nichol and published by Mill can be found in: John P. Nichol, 'ART. I.-On Political Economy, in connexion with the Moral State and Moral Prospects of Society.', *Westminster Review*, Jul 1832; 17, 33: 1–33.

[47] Herschel to Mill, 16 July 1845, Royal Society Herschel MSS HS 12.332.

[48] John F. W. Herschel, 'Address to the British Association for the Advancement of Science, June 19. 1845' in *Essays from the Edinburgh and Quarterly Reviews, with Addresses and Other Pieces* (London: Longman, Brown, Green, Longmans & Roberts, 1857), 634–682 at 666–667.

At the same time, there must be a distinct suspicion that Herschel's vehement rejection of Comte's 'proof' of the nebular hypothesis also related to its use by Chambers in the opening chapter of *Vestiges*. As Mill subsequently confided in a letter to the French positivist:

> Or, l'auteur des "Vestiges" cita votre vérification de l'hypothèse de Laplace. C'est, je crois, ce qui détermina l'attaque de Herschel, et Sedgwick vous attaque de même dans le passage ci-joint.[49]

Wanting to maintain his reputation as an objective arbiter and possessing an interpersonal style which stemmed from his shyness and avoidance of conflict, Herschel asked Mill to try not to give offence to the French humanist as a result of his remarks.[50]

Herschel was then, by dint of his reputation and position in society, an experienced astronomer and a widely accepted referee who had adopted an intermediate position on the reliability of the nebular hypothesis. His well-known views on the merits of inductive thinking, when combined with the earlier comments made by Whewell on the potential for future proofs of the nebular hypothesis, established an intellectual environment in which there was space for the production of new empirical evidence: an opening which Rosse and Robinson initially believed they had filled by means of an endeavour of industrial proportions.[51]

The Emergence of Doubt

Although the question of the nebular hypothesis had apparently been 'resolved' by the use of the Leviathan, Rosse in 1850, following the tragic events of his daughter's death and the potato famine, admitted to distinct doubts about the objectivity of the observations he had previously reported:

[49] Frances E. Mineka (ed.), *The Earlier Letters of John Stuart Mill 1812–1848 (Volume XIII of the Collected Works of John Stuart Mill)*, (University of Toronto Press, 1963), 677–679, Letter 467, to Auguste Comte, dated 18 July 1845.

[50] See Schweber, 'Herschel', 16–17, for comments on Herschel's shyness and avoidance of conflict. In Herschel to Mill, 13 July 1845, Royal Society Herschel MSS HS 12.330, Herschel asks Mill to avoid upsetting Comte. The salient portion of this letter is also reproduced in Mineka, *Letters, Volume XIII*, 675 n3.

[51] On the popularity of Herschel's manual of inductive reasoning, see: James A. Secord, *Visions of Science: Books and Readers at the Dawn of the Victorian Age* (Oxford University Press, 2014), 80–106 at 80–82.

> When certain phenomena can only be seen with great difficulty, the eye may imperceptibly be in some degree influenced by the mind; therefore a preconceived theory may mislead, and speculations are not without danger.[52]

Rosse's evident uncertainty about the validity of his results, which may also reflect a revised take on Robinson's "preconceived theory", was in due course reinforced by the comments of some of his peers. In 1851, for example, Sir George Biddell Airy (1801–1892), the Astronomer Royal, told the British Association that "he denied the inference that 'all nebulae are resolvable'", and in 1853 the experienced German observer Otto Wilhelm von Struve (1819–1905), a professional astronomer who was the director of the observatory at Pulkovo in Russia, expressed similarly severe reservations: "the alleged miracles of resolution are nothing but illusions."[53]

Robinson, for his part, was satisfied, at least initially, in both private and public with the outcomes of his observations of nebulae through the eyepiece of the Leviathan: "We have however worked to some purpose, for of the 43 nebulae which we have examined All have been resolved."[54] It is interesting to note the subtle contrast here between the parallel claims of Rosse and Robinson; the dinner conversation of the former made it clear that at least five nebulae remained unresolved.[55]

Within a few years of the original observations, however, Robinson's confidence in the correctness of his conclusions had also decreased, particularly given that he must have been sufficiently self-aware to be conscious that some of his reports of resolution were, to say the least, somewhat premature. In 1849, at a time when he might have hoped that the nebular hypothesis had been fatally holed below the waterline, his correspondence with John Herschel, whose views he had previously denigrated, revealed his own uncertainty as to whether changes in the shapes of nebulae occur over time (a potentially significant argument against the

[52] Third Earl of Rosse, 'Observations on the nebulae' in Charles Parsons (ed.), *The Scientific Papers of William Parsons, Third Earl of Rosse 1800–1867* (Cambridge University Press, 2011), 109–124 at 113–114. It is now clear to astronomers that the nebula is a mixture of stars and gas and cannot have been 'resolvable' in the sense inferred by Rosse et al.

[53] Schaffer, 'Leviathan', 218. The embedded quotes are from: George B. Airy, 'Address', *British Association Report* (1851), xxxix–liii at xli and Michael A. Hoskin, *Stellar Astronomy: Historical Studies*, (Chalfont St. Giles: Science History Publications, 1982), 150.

[54] Robinson, *Astronomical Diary*, 12 March 1845, Birr Castle MSS L 2.1.

[55] Rosse apparently made a slightly more restricted claim that 42 out of 47 nebulae had been resolved. See: Frances Power Cobbe to Louise Cobbe, dated to the 1840s, Birr Castle MSS J 7.

hypothesis that all nebulae are clusters of stars and that the universe has been static or has exhibited only uniform motion since its creation):

> From what I have seen of nebulae, I am the more anxious that no time should be lost in securing accurate drawings, based on exact measurements, of those that are most remarkable among them. Whatever changes occur in them can only thus be ascertained; and it seems desirable that the survey be made with instrumental means as powerful as the existing condition of optical art is able to furnish.[56]

Herschel had, as early as 1826, commented on potential changes in the appearance of the Orion nebula. Given his tendency to avoid courting controversy or entering into contentious topics of discourse, however, Herschel characteristically deflected Robinson's request by concurring that he looked forward to the time when his own observations of the southern heavens could indeed be repeated.[57]

A striking feature of Robinson's request is not only the tone of anxiety he evinced but also the implicit trust which the director of the Armagh Observatory was prepared to place in the products of equipment other than the Leviathan and a potential reliance on the authority of Herschel, whose accuracy in the production of drawings and of evidence for the changing shape of the Orion nebula had been highlighted by Airy in 1836.[58] This newfound uncertainty contrasted starkly with Robinson's earlier comments to a Dublin audience in which he had favourably compared the openness with which the details of the design and manufacturing processes undertaken by Rosse had been documented with the lack of information on telescope manufacturing provided by William Herschel.[59] The alleged shortfall on the part of the elder Herschel was viewed by Robinson as evidence pointing to the substandard performance of Herschel's 40-foot reflector: on which subject he conducted a brief letter-writing campaign in 1843.[60]

[56] Robinson to Herschel, 21 October 1849, Royal Society Herschel MSS HS 14.375.

[57] Buttmann, *Telescope*, 50. Herschel to Robinson, 27 October 1849, Royal Society Herschel MSS HS 14.376.

[58] Schaffer, 'Leviathan', 204.

[59] See, for example, Robinson's presentations to the Royal Irish Academy in 1840 and 1845 reprinted in Parsons, *scientific*, 14–31. The number of pages devoted to the description of the construction of the telescope are significantly greater than those which report actual findings and observations.

[60] Thomas R. Robinson, 'Dr. Robinson's reply to Sir John Herschel', *The Athenaeum*, Oct 21, 1843; 834: 945–946 and 'Dr. Robinson's reply to Sir John Herschel', *The Athenaeum*, November 18, 1843; 838: 1027–1028.

As Simon Schaffer has pointed out, this attack on the reliability and quality of Herschel's equipment was deployed to highlight important differences between the two telescopic teams, including the far more open approach to design documentation and description of the construction processes delivered by the Parsonstown team. These differences align with the contrasting class backgrounds of the two rival constructors. Rosse was an aristocratic amateur demonstrating his grasp of technology and his technocratic command of his workforce. William Herschel, by way of contrast, had started his working life as a bandsman and had for much of his career acted in essence as a craft worker, creating a cottage industry of telescope production in which he of necessity preserved the secrets of his success to maintain the profitability of his enterprise.[61]

Robinson's earlier attack on the reliability of William Herschel's equipment implied an indirect assault on the reliability of his nebular observations and was therefore also an animadversion against the latter's theory of the nebular genesis of stellar systems. Faced with this challenge, John Herschel supported his father's legacy and attempted to defend him against the charge that the largest telescope constructed by his father in Slough was ineffectual.[62] It is more than a little ironic, therefore, that in the passage referring to possible changes in the Orion nebula quoted above, Robinson had begun to appreciate the limitations of the Leviathan and had come to accept that the Birr Castle telescope did not necessarily represent the "instrumental means as powerful as the existing condition of optical art is able to furnish".[63]

Despite the animosity of some of Robinson's interactions with and comments about John Herschel in the early 1840s, it is apparent that he was viewed as the great authority of the day on sidereal and nebular observations, and his views carried great weight with the Irish 'grand amateurs' as well as the wider scientific establishment.[64] Herschel understood that there was, in fact, a clear demarcation in terms of both scale and implication for

[61] Schaffer, 'Leviathan', 190–191.

[62] John F. W. Herschel, 'Sir John Herschel's reply to Dr. Robinson', *The Athenaeum*; November 4, 1843; 836: 983–984.

[63] Robinson to Herschel, 21 October 1849, Royal Society Herschel MSS HS 14.375.

[64] For a discussion of the Irish 'grand amateurs' and their astronomical work, see, for example, Allan Chapman, 'William Parsons and the Irish nineteenth-century tradition of independent astronomical research' in Charles Mollan (ed.), *William Parsons 3rd Earl of Rosse: Astronomy and the Castle in Nineteenth-Century Ireland* (Manchester University Press, 2014), 271–297.

the two concepts contained within the broad term 'nebular hypothesis'. On the one hand, he was unwilling to publicly contradict the authority of Lord Rosse's gentlemanly claims of the potential resolvability of many nebulae, whilst on the other he was willing to retain his father's view that many objects in the universe are not formed solely from stars. He was somewhat sceptical about the verifiability and the dynamics of the Laplacian version of the 'nebulous hypothesis', but nonetheless appeared to keep an open mind on the topic. His speech to the British Association at Cambridge in 1845, for example, reflected genuine uncertainty as to where the 'truth' then lay.[65]

The Stellar Universe

Nichol, on the other hand and on the basis of his earlier arguments, was happy to present the Laplacian component of the nebular hypothesis as a given and was ready to move on to the consequences of his assumption. In 1848, he published another popular astronomy text entitled *The Stellar Universe*, which was intended to be a more affordable rehash of his previous works aimed particularly at "younger Intelligences", by which he seems to have meant both children and women.[66] *Tait's Edinburgh Magazine* concurred with this view, describing the work as being of "*mignon* prettiness" which "is perhaps, however, more a lady's than a gentleman's' book and may make many Mrs. Somervilles".[67]

This publication, aimed at a wider audience than any of his previous output, and in keeping with the juvenile nature of at least part of the intended audience for both his lectures and his books, contained a rare exception to his normal theological vagueness in the form of an apparent nod towards conventional religion:

> God being known by the immediate insight of him 'who doeth the will of the Father;' God being known, of the child of God, in and through the affiliative act of genuine worship, the unfathomable and resplendent abyss of Space in Time is simultaneously converted into an illuminated commentary upon the transcending attributes of Jehovah.[68]

[65] John F. W. Herschel, 'Address to the British Association for the Advancement of Science, June 19. 1845' in *Essays from the Edinburgh and Quarterly Reviews, with addresses and other pieces* (London: Longman, Brown, Green, Longmans & Roberts, 1857), 634–682 at 659–668.

[66] Nichol, *Stellar*, viii.

[67] 'Professor Nichol's new works', *Tait's Edinburgh Magazine*, March 1848; 15, 171: 200.

[68] Nichol, *Stellar*, 226.

His approach here seems deliberately ambiguous. References to "him 'who doeth the will of the father'" and "the child of God" would be read by many of his audience as being Trinitarian but can alternately be understood as terms which describe ordinary human beings.[69]

Stellar Universe was in fact somewhat of a wolf in sheep's clothing, as Nichol made full use of the opportunity it provided to revisit his earlier calculations, based on the seeing power of 'Lord Rosse's telescope', to further expand both astronomical and geological timescales. To emphasise the great age implied by this update, he reintroduced an idea which he had earlier credited to John Herschel.[70] He invited the readers, as stellar observers, to contemplate a novel cosmological point of view in which they came to understand that the arrangement of constellations in the sky has not been fixed since the dawn of time, a natural assumption based on the limited duration of a single lifetime or even the entirety of documented astronomy, but must have changed gradually due to the solar system's motion through space:

> Free, then, from fear, let us rise far higher; and assuming that Bessel's Star indicates the average distance of the nearest Orbs, we infer the Sun would require no more than 500,000 years to reach that extremest verge at which the eye can descry a single Star; nay, it could reach that remotest distance to which Lord Rosse's telescope can pierce, in about TWO HUNDRED AND FIFTY MILLIONS OF YEARS; and so far is even this stupendous period from sounding all the working time of nature, that many of the mountains of our Earth may, through its whole duration, have been in being, rearing their peaks towards different constellations, and surviving, in their littleness and fragility, even these immense transitions![71]

The idea that the appearance of the constellations would change and evolve over very long periods of time was a highly affective trope which found further employment in H. G. Wells' *The Time Machine*:

> Above me shone the stars, for the night was very clear. I felt a certain sense of friendly comfort in their twinkling. All the old constellations had gone

[69] Nichol has paraphrased Matthew 7:21, substituting the word "the" for the word "my" in the phrase "who doeth the will of my Father", thus turning a reported saying of Jesus into a more generic exhortation.

[70] Nichol, *Thoughts*, 180.

[71] Nichol, *Stellar*, 196–197. Emphasis in original.

> from the sky, however: that slow movement which is imperceptible in a hundred human lifetimes, had long since rearranged them in unfamiliar groupings.[72]

This implication of the potential consequences of deep time was no doubt welcome to contemporary geologists whilst at the same time recalibrating the chronological 'set points' of his audience and promoting the likelihood of the cosmic timescales required by developmental hypotheses in general.

Opiate Addiction

By this time, Nichol had succumbed to the influence of a cocktail of pharmaceuticals which he later reported as a "long standing" dependency on "opiates and anodynes".[73] In his description of the pleasure which he experienced during the earlier phases of opium dependency, he identified several benefits of the drug including freedom from pain combined with feelings of physical and mental potentiation:

> And it endowed me with such strength besides,—power alike muscular and intellectual. Prostrate, as I might be, racked by pain, mind and body exhausted and prone—hand me that draught [of opium], and in one quarter of an hour I could face any danger, and contend with all difficulty.[74]

The consumption of these psychotropic substances had initially also delivered a sense of *joie de vivre*, "a delicious feeling of pleasure in mere *being*, which the late metaphysician, Dr Thomas Brown considered a separate and distinctive *sense*".[75] In discussing the latter affect, he was making, in passing, a mental genuflection towards the Edinburgh-based philosopher Thomas Brown (1778–1820), who, as Thomas Dixon has documented, was one of the founding fathers of psychology, being in particular responsible for introducing the concept of emotions as a separate analytical category and whose series of *Lectures on the Philosophy of the Human*

[72] Herbert G. Wells, *The Time Machine: An Invention* (London: Heinemann, 1895), 143–144.

[73] John P. Nichol, *Memorials from Ben Rhydding: Concerning the Place, Its People, Its Cures* (London: Charles Gilpin, 1852), 2–3.

[74] Ibid., 4.

[75] Nichol, *Ben Rhydding*, 4.

Mind were published posthumously in 1820. Brown's descriptions of a "designing Power" which reflected his "preference for natural over revealed religion" and his advocacy "of the separation of academic posts from religious requirement" can be seen to foreshadow Nichol's later pronouncements on both of these topics.[76]

The early positive effects of opium were, however, eventually replaced by far-less-pleasant visions which caused him to suffer significant sleep loss:

> But the day of vengeance—of terrible although most fitting compensation—came swift [...] And through an appalling kind of propriety, the blow fell earliest on that very portion of my existence which was first soothed by the hollow promise of consolation—I mean the hours of sleep.[77]

In due course, Nichol travelled to Ben Rhydding in Yorkshire to take a course of hydrotherapy as a potential cure for his opiate addiction. It is not clear from Nichol's account, published in 1852, when he had first been exposed to opiates but it is certainly possible in the general context of early nineteenth-century pharmaceutical treatment options, that they had been prescribed during the earlier bout of the "long standing" illness which permeated his life and, in particular, preceded his theological breakdown.[78] In any case, it is evident that his opiate dependency had taken a distinct turn for the worse "five years ago" when he was "stricken by a low and nervous fever, the issue of profound and protracted anxieties". His "physician, than whom a kinder never existed, fondled [him] with opiates".[79] We can therefore safely conclude that for most of the period during which he enjoyed his greatest successes in publishing and lecturing, he would have also been regularly consuming psychoactive agents. His penchant for psychotropic medication and his sporadic ventures into unconventional sciences reflect a tendency, shared with his friend Thomas De Quincey (1785–1859), to experiment with novelty and affect.

[76] Thomas Dixon, *From Passions to Emotions: The Creation of a Secular Psychological Category* (Cambridge University Press, 2003), 109–127, esp. 113–114.

[77] Nichol, *Ben Rhydding*, 5.

[78] Virginia Berridge, *Demons: Our Changing Attitudes to Alcohol, Tobacco, & Drugs* (Oxford University Press, 2013), 12–35. Nichol, *Ben Rhydding*, 2.

[79] Nichol, *Ben Rhydding*, 3.

Support from William Whewell and Herbert Spencer

At this juncture, it is appropriate to briefly revisit the evolving views of William Whewell (1794–1866), the man who originally coined the term 'nebular hypothesis' and who was now to almost inadvertently provide some support for Nichol's point of view. In 1853, Whewell anonymously published *Of the Plurality of Worlds: An Essay*, in which, for the reasons related in Chap. 4, he expressed a number of theological and cosmological doubts concerning the possibility of life on other planets. In one of his supporting arguments, he deployed an uncritical acceptance of the gaseous nature of nebulae, based largely on John Herschel's observations of the heterogeneous structure of the Magellanic Clouds, which can be seen to contain both stars and nebulae lying side by side, thus apparently disposing of the idea that nebulae are in fact clusters of stars which might in turn play host to extra-solar planets.[80] In adopting this new point of view, Whewell was subtly negating the observations made at Birr Castle. Without anywhere overtly stating that Lord Rosse was wrong, he nonetheless proceeded to accept one leg of Nichol's fork and to claim that the nebulae are in fact "vast masses of incoherent or gaseous matter, of immense tenuity".[81] Whewell had, for his own purposes, managed to reshape a belief in Tennyson's "fluid haze" into support for a form of traditional religion. In the mid-nineteenth century, the evidence for or against either of the nebular hypotheses was quite malleable in the minds of both observers and commentators.

Despite his repeated illnesses and heavy consumption of opiates, Nichol survived until September 1859, long enough to have appreciated both Whewell's unsolicited, almost inadvertent, support as well as a spirited defence of the nebular hypothesis and its applicability to planetary development by the philosopher and early sociologist, Herbert Spencer (1820–1903). Spencer was a fellow sufferer from a series of ailments, including insomnia and dyspepsia, which, as with Nichol's recurring illness, combined manifest physical symptoms with hard-to-identify causes. His writings, especially his autobiography, frequently revealed his anxiety about his fluctuating state of health as well as a recurrent fear that he would at some point

[80] Michael J. Crowe, *The Extraterrestrial Life Debate 1750–1900: The Idea of a Plurality of Worlds from Kant to Lowell* (Mineola NY: Dover Publications, 1999), 285–286. For further discussion, see Chap. 2.

[81] William Whewell, *Of the Plurality of Worlds: An Essay* (London: John W. Parker and Son, 1853), 138.

succumb to overwork in much the same way as had, in his view, various relatives including his father.[82] At times, like Nichol, Spencer resorted to opium in various forms as a temporary solution for his insomnia:

> This mention of opium reminds me that I had for some time previously made occasional use of it—commonly under the form of morphia. With me sleep brought sleep and wakefulness was habitually followed by more wakefulness; so that after a series of specially bad nights it had been my practice to break the morbid habit, and re-establish the periodicity of sleep by artificial means. Sometimes it was weeks, sometimes months, before I again had recourse to one or other preparation of opium. That the average result was beneficial is an opinion which I here express, because there is, I think, an undue fear of opium; both in the minds of medical men and in those of men at large.[83]

The self-identified causes of Spencer's recurrent sleeplessness were many and various, including "making blood faster than my brain can bear", "*defect* [shortage] of blood in the brain" and that "sea air has been doing me harm rather than good". "Among the remedies recommended by his friends at Standish [a Mr. and Mrs. Potter] were marriage and the exercise of the emotions, more especially the religious emotions." Part of Spencer's response to these suggestions, which was excised from the version of a letter to Mrs Potter which was published in his autobiography, but was then reinstated by his biographer, David Duncan, provides an interesting insight into those elements of his early upbringing which led to feelings of aversion to the forms and content of traditional religion:

> Owing to the foolish pertinacity with which, as a child, I was weekly surfeited with religious teachings and observances, I have contracted a decided repugnance to the very forms in which they were conveyed. I cannot hear scriptural expressions without experiencing a certain disagreeable feeling; and I can no more escape this than I can the nausea produced in me by particular sweets that were commonly given me after medicine when a child.

[82] Herbert Spencer, *An Autobiography*, vol I (London: Williams and Norgate, 1904), 31–32, 58, 66.

[83] Herbert Spencer, *An Autobiography*, vol II (London: Williams and Norgate, 1904), 92–93.

[...] And further, it is not as though your plan had never been tried. Up to seventeen I was constantly in the way of hearing the gospels.[84]

The negative psychotropic effect of his frequent exposure to nonconformist theologies in his youth, "habitually" attending a Quaker meeting with his father and a Methodist service with his mother on the same day, goes a long way to explaining his subsequent disenchantment with organised religion.[85] Spencer's later metaphysical and theological positions can be summarised as a belief that whilst many, if not all, aspects of natural law can be discovered by man, some key elements of the universe will forever remain unknown and which he therefore attributed to an undefined prime mover whom he labelled 'The Unknowable'.[86]

Like Proctor, a fellow believer in The Unknowable, Spencer also built on the Darwinian hypothesis of biological development and averred that the universe as a whole was subject to a "law of organic evolution". When discussing his idea of universal progress in *First Principles of a System of Philosophy* (1865), Spencer suggested that the nebular hypothesis constituted an important proof point for the validity of evolutionary processes acting at all scales from the macroscopic changes associated with cosmogony down to the minutiae of interactions within human societies:

Now I propose in the first place to show, that this law of organic evolution is the law of all evolution. [...] From the earliest traceable cosmical changes down to the latest results of civilization, we shall find that the transfer of the homogeneous into the heterogenous, is that in which Evolution essentially consists. [...] *if* the Nebular hypothesis be true, the genesis of the solar system supplies one illustration of this law.[87]

Prior to the original publication of *First Principles*, Spencer had already composed a lengthy apologetic for the nebular hypothesis which was

[84] David Duncan, *The Life and Letters of Herbert Spencer* (London: Methuen, 1908), 79–80. Duncan reproduced portions of a series of letters written by Spencer to his father on 29 September, 10 October, 20 October and 30 October 1855 in which he made various suggestions as to the causes of his insomnia. The comments about his reaction to religion are contained in a portion of a letter written by Spencer to Mrs. Potter in early 1856. The remainder of this letter was published by Spencer himself in: Spencer, *Autobiography*, vol I, 478–479. The Potters' daughter Martha Harriet (1858–1943) was the future wife of Sidney James Webb (1859–1947).

[85] Ibid., 83.

[86] See, for example, Herbert Spencer, *First Principles of a New System of Philosophy* (New York: D. Appleton, 1865), 98–123.

[87] Ibid., 148–149. See discussion of Proctor's religious and metaphysical views in Chap. 5.

published in the *Westminster Review* of July 1858.[88] In this article, Spencer, like Nichol, did not directly dispute the Birr Castle observations of the Orion nebula, they being founded on the authority of an aristocrat. He positioned them instead as potentially irrelevant on the grounds that sidereal development within nebulae may now have ceased throughout the universe, and he then restated the planetary cosmogony of Laplace in his own words. One of his more speculative points, the supposition that the earth's core must be gaseous (to explain the observed rate of precession), had to be retracted in a subsequent (1865) reprint of the article. A few other calculations were found to be based on faulty assumptions but on the whole his thesis stood the test of time rather better than might have been expected given that, as with many of his frequent and prolix interventions in diversified topics, his approach was founded on a limited theoretical background.[89]

In the 1865 reprint, Spencer was seemingly unaware of the recent discovery by William Huggins (1824–1910) of the gaseous state of the nebula in the constellation of *Draco*, which was assumed by many to demonstrate the existence of William Herschel's 'shining fluid' and to be clinching evidence in favour of the nebular hypothesis.[90] Even the Birr Castle team accepted that spectroscopy had swung the argument decisively towards the existence of true nebulosity and devoted some of their telescope time to attempts to capture the spectra of additional non-stellar objects.[91] As a consequence of this shift to general acceptance of nebular cosmogony, an 1890 review of a further reprint of Spencer's article was willing to concede that the passage of time had increased the acceptability of his arguments and that further observations had provided significant evidence in favour of the "nebular hypothesis, or theory as it is now more commonly called".[92]

[88] Herbert Spencer, 'Recent astronomy, and the nebular hypothesis', *Westminster Review*, July 1858; 70, 137: 185–225.

[89] Herbert Spencer, 'Recent astronomy, and the nebular hypothesis', *Illustrations of Universal Progress; a Series of Discussions* (New York: D. Appleton, 1865), 239–299. See, in particular, the footnotes on pages 275, 279 and 291.

[90] Based on spectroscopic evidence. See Anna Henchman, '"The Globe we groan in": Astronomical Distance and Stellar Decay in In Memoriam' *Victorian Poetry*, 41, 1, Spring 2003: 29–45 at 36. Also see Agnes M. Clerke, *A Popular History of Astronomy during the Nineteenth Century*, 2nd Edition (Edinburgh: A. & C. Black, 1887), 437, for her vindication of Herschel.

[91] Wolfgang Steinicke, 'Birr Castle observations of non-stellar objects and the development of nebular theories' in Mollan (ed.), *Parsons*, 210–270 at 259–260.

[92] 'Astronomical Notes', *The Athenaeum*; 1 March 1890; 3253: 279.

The Elusiveness of Objectivity in the Making of Cosmogonies

These three chapters have primarily examined some of the contextual and cultural influences which affected the main participants in the nebular hypothesis debate in order to study, in particular, the *Weltmodell* changes which integrated affect and cognition to influence their selections of epistemological positions. Perhaps the most obvious and unsurprising conclusion which suggests itself is the evident correlation between the theological and political avocations of key participants and their attitudes to the topic of nebular cosmogony. The battle-lines between a static, conservative universe and the more progressive, evolutionary alternative are much clearer than was the case with pluralism, where a belief in extraterrestrial plenitude was capable of being deployed both for and against traditional theology.

In this context, it is relevant that the two individuals who lived in the heterogeneous and contested religious environment of Ireland appeared to be the most wedded to traditional theology and cosmology. Thomas Romney Robinson and Lord Rosse, in different ways, experienced significant difficulties in their relationships with the other inhabitants of the island, either in managing religious dissent, political opposition, famine and unemployment (Rosse) or in the confrontations engendered by the process of collecting tithes from his parishioners (Robinson).

Rosse made very few statements about his personal religious beliefs. He engaged in extended projects of telescope construction, inspired by his father's interest in nebular cosmology: an interest which had become evident after the early death of his younger brother and the emergence of his father's consolation-seeking invocation of the evidence of astronomy in support of the truth of theism.

Although the Birr telescopes began life with a fairly overt theological purpose, that of negating the nebular hypothesis, this objective subsequently mutated, at least on the part of Rosse, into a desire to evince mastery of the required technologies by means of demonstrations of their effectiveness. The somewhat-doubtful resolutions of nebulae were supported by the promotion of an illustration of the enigmatic spiral nebula M.51, which had been produced as a result of the superior light collecting powers of the Leviathan and which could be favourably compared with the earlier engravings of the same object made by John Herschel. As such,

M.51 and subsequent spiral nebulae were adopted as the best proofs of the technical power of the giant reflecting telescope.[93]

The key person who was responsible for the original mission of the Leviathan was Robinson. As a younger man, he had, like Nichol, demonstrated a nascent literary capability which was informed by his Ulster Episcopalian background. This combination of *Weltmodell* influences produced an amateur astronomer able and willing to deliver expansive sermons inflected by hyperbole. Robinson viewed Rosse as an aristocratic sponsor, encouraging the peer to invest large amounts of his wife's money into a potentially nugatory project. In return, he acted as Lord Rosse's mouthpiece when speaking to the Royal Irish Academy: speeches which were subsequently transformed into published papers in which he lavishly praised his patron's technical abilities whilst attempting to deprecate the prior truth claims of the Herschels.

Any potential aspiration to mechanical objectivity was for Robinson epistemologically subservient to the truths revealed by religion, and so he predictably succeeded in interpreting the available evidence from Lord Rosse's telescopes as support for his expected conclusions. The nebular hypothesis was, in Robinson's mind, an attack on the church as evidenced by its use in the first section of *Vestiges*. Once he felt that he had achieved the desired results, however, he began to entertain doubts about his own evidence and the 'truth' of his claims.

Like Robinson, John Pringle Nichol had developed a taste for poetic self-expression during his childhood in Scotland, fuelled by a fertile imagination: an attribute which was rewarded by the positive responses he received from his peers and the public. Unlike Robinson, however, he was exposed to religiously heterodox ideas in his early years. Nichol was also a political radical and, following the severe illness which caused a significant shift in his *Weltmodell* and his resignation from the established church, evidently continued to hold unorthodox views on both religion and politics into later life.

By 1850, Nichol had succeeded in carving out a niche as a preacher of the new religion of astronomy, and the political inclinations of the older man were perhaps sated, or at least blunted, by the recent successes of other would-be reformers in Britain, which could be contrasted with the repetition of the dangers of revolutionary fervour represented by recent

[93] Trevor C. Weekes, 'The Nineteenth-Century Spiral Nebula Whodunit', *Physics in Perspective*, 12 (2010): 146–162.

uprisings in France and elsewhere in Europe. Nichol also developed a dependency on a cocktail of opiates and other drugs at about this time. It is interesting to note the potential relationship between his success as a public speaker and the onset of his consumption of these psychotropic substances which he described as producing a state of ecstatic passivity "neither dreamful nor yet wholly unconscious".[94]

The final key point which can fruitfully be extracted from the history of the nebular hypothesis is the elusiveness of objectivity in the making of cosmogonies. Robinson and Nichol both managed the gathering of data and the deployment of argumentation in ways which bolstered their preferred metaphysical and theological positions. Robinson had by 1840, well before the completion of the Leviathan, convinced himself that the nebulae were all going to be resolvable. He was thus easily able to interpret subsequent observational evidence as support for his diagnosis. Such was his enthusiasm that, even in his absence, Lord Rosse was almost able to resolve the great nebula in Orion.[95] Nichol, faced with aristocratic certainty and the potential embarrassment of his unauthorised contributions of *materiel* for a project for which he had held out rather different hopes, nonetheless managed to snatch a certain kind of victory from the jaws of potential defeat by his subsequent reinterpretation of claims of resolution and thence the promotion of an expanding and progressive universe.

[94] Nichol, *Ben Rhydding*, 4.

[95] Robinson's "publicity campaign" was also successful in persuading at least one other astronomer to 'resolve' the Orion nebula as proof of the efficacy of his own telescope. See: Schaffer, 'nebular', 140.

PART IV

The Ages of the Earth and Sun

CHAPTER 9

"And Murmurs from the Dying Sun": William Thomson, Thermodynamics and Theology

O Sorrow, cruel fellowship,
O Priestess in the vaults of Death,
O sweet and bitter in a breath,
What whispers from thy lying lip?
'The stars,' she whispers, 'blindly run;
A web is wov'n across the sky;
From out our waste places comes a cry,
And murmurs from the dying sun:

Alfred Lord Tennyson, '*In Memoriam A. H. H.*', III[1]

Fifty-one years after the near-fatal malady which afflicted Thomas Chalmers (1780–1847), recounted at the beginning of Chap. 3, another Scot experienced an extended period of recuperation during which he was confined to a bedroom for many weeks and treated with psychotropic substances to dull the pain. On this occasion, however, the man in question was not a cleric by profession and his injury was not necessarily life threatening. The patient was the Glaswegian academic William Thomson (1824–1907, later Lord Kelvin), who had already earned a considerable reputation in a variety of fields associated with astronomy and the physical sciences, including being credited with having played a significant role in the

[1] Alfred Lord Tennyson, 'In Memoriam A. H. H.' in Karen Hodder (ed.), *The Works of Alfred Lord Tennyson* (Ware: Wordsworth Editions, 1994), 305–389 at 311.

H. Carlton, *Cosmology and the Scientific Self in the Nineteenth Century*, https://doi.org/10.1007/978-3-031-05280-4_9

discovery of the laws of thermodynamics. He had also been profitably engaged in such practical engineering matters as the laying of submarine telegraph cables.

One Saturday in December 1860, whilst staying with friends on the west coast of Scotland, he had decided to engage in his favourite "pastime of 'curling' on the ice". In an unfortunate accident, he fell in such a way and experienced so much pain that it appeared he might have fractured his leg. He had indeed broken the bone, but this important diagnosis was not delivered until the following Friday and only after the engagement of a second opinion. The leg was then reset successfully, but the pain which he suffered in the intervening period had been considerable and as a consequence Thomson had been treated with "continual opiates".[2]

His wife Margaret (née Crum, c.1830–1870) was particularly worried about the possible long-term consequences of the injury to his left leg. "It was found to be an inch and a half shorter than the other, and had to be stretched out to the same length. Our chief anxiety is as to whether bad effects will be left." His recovery from this distressing event was very slow. "I begin to fear that he will not be able to return to his work this Session. He eats and sleeps pretty well, but is exceedingly thin, and is easily fatigued by any mental exertion."[3] Thomson had initially hoped to return to work at the University of Glasgow towards the end of the first quarter of 1861. But by the time March arrived, it was evident that his mobility was still considerably impaired and that the recovery process was going to run into several more months. "The Dr thinks it will be a considerable time before I can get up & down stairs, & says I must give up the hope of getting to work before the end of the session."[4]

Because of the delay in setting the bone, his wife's fears for his future mobility were also duly realised, and he was left with a shortened leg which caused him to walk with a distinctive limp for the rest of his life so that he lost the ability to undertake the strenuous walks which he had previously enjoyed according to biographer Silvanus Phillips Thompson (1851–1916):

[2] Silvanus P. Thompson, *The Life of William Thomson: Baron Kelvin of Largs*, vol I (London: Macmillan, 1910), 411–412.

[3] Margaret Thomson to Hermann von Helmholtz, 11 January 1861, in Thompson, *Life*, vol 1, 412–414.

[4] William Thomson to Elizabeth King (his sister), 28 January 1861, in Thompson, *Life*, vol 1, 415–416. William Thomson to James David Forbes, 23 March 1861, Courtesy of the University of St Andrews Libraries and Museums, Special Collections, msdep7, 1861, letter 43.

> The injury left him permanently lame. He used to walk fast, limping along with his left hand pressed against the hip. When in the summer they returned to their favourite resort at Kilmichael, in Glen Cloy, Arran, he wistfully remarked that he would never again be able to climb Goatfell and he never did.[5]

Thomson did possess, both before and after the accident, a devout but relatively non-doctrinal religious belief, regularly attending either Anglican or Presbyterian church services depending on his location. His niece, Agnes Gardner King (1856–1929), reported that the future peer was a firm believer in the principle of 'Intelligent Design' and that:

> All through life Lord Kelvin [...] had a simple, childlike faith which pervaded his whole being. The deeper he delved into Science and the more he studied its mysteries, the greater his veneration for the Maker of all.[6]

Thomson's *Weltmodell* must have been significantly impacted during the painful episode which caused him to be laid up in Largs for a lengthy period.[7] Unlike the account composed by William Hanna (1808–1882) which presented his father-in-law's period of convalescence in the form of a conversion narrative, featuring Chalmers as a latter-day apostle Paul, there is no parallel narrative of Thomson having 'seen the light' or of having experienced a significant shift in his theological thinking during this time, despite the severity of the injury and the permanence of its after-effects, which left him with plenty of time to reflect on the brevity of life and the entropic 'running down of the clock'. His serious injury and long recuperation would have been an affective demonstration that the dissipation of the physical universe which he had portrayed in his thermodynamics also applied to biological systems. Which, of course, had degenerative and ultimately fatal consequences for all living things including himself. He was well aware that time does not extend indefinitely for either heavenly or human bodies and indeed he concluded that it is insufficiently extensive to meet the needs of uniformitarian geologists and biological

[5] Thompson, *Life*, vol 1, 416–417.

[6] Agnes G. King, *Kelvin the Man* (London: Hodder and Stoughton, 1925), 29, 99, 28. Agnes was Thomson's niece.

[7] A German word meaning 'world picture' which has previously been employed to describe both the cosmological models of physical scientists and the creation stories found in traditional religions. It thus encompasses the two key epistemologies under discussion here.

evolutionists. As we shall shortly see, Thomson's post 1860 theoretical work exhibited a marked increase in thermodynamical attacks on the extended timescales postulated by Darwin and others as well as some speculations about the nature of free will.

The North British Physicists

Before we continue the examination of William Thomson's intervention in the discussion of evolutionary timescales, we should first explore the shared cultural inputs and the religiously informed backgrounds which have been argued as having been influential in shaping the views of the physicists with whom he is associated and who are collectively referred to as the 'North British group'. Their shared desire to refute 'long' chronologies which ran counter to their theological inclinations was certainly one theme which connected this otherwise loosely bound collective, particularly during the 1860s and 1870s.

A second key thread which similarly had religious connotations was the question of where in the universe, or even outside the universe as we know it, is the locus of those supposed components of sentience which we refer to as free will, consciousness and the soul. This question must be set against the background of the contemporary vogue for materialistic explanations of the invisible phenomena which underlie the visible universe. For a scientist who wished to retain a role in the world for a theistic deity, there had to be both a space and a time in which such an entity could continue to intervene in human destiny. And if the mind, acting as a proxy for the soul, has an existence separate from the physical body, then it also had to have a location and a mode of operation which allowed it to interact with the physical world. The rise of spiritualism and other heterodox forms of belief can be seen to have been a parallel response to scientific naturalism at a time when steam-driven machinery and the electric telegraph could be read as being analogous with the musculoskeletal and nervous systems of the human body.[8]

In addition to the relevant interventions made by William Thomson, we shall also be examining the contributions of four other scientists who are often cited as fellow members of the North British group. They include his colleague and occasional co-author Peter Guthrie Tait (1831–1901), the theoretical physicist James Clerk Maxwell (1831–1879), the

[8] See Chap. 1 for a discussion of theological trends in the later nineteenth century.

astronomer and physicist Balfour Stewart (1828–1887) and Henry Charles Fleeming Jenkin (1833–1885). Jenkin, in paricular, can reasonably be regarded as a true polymath, having variously functioned as a physicist, an electrical engineer, literary critic and political economist.[9]

Before proceeding to examine in detail the experiential journeys undertaken by these subjects, it is worth pausing to revisit what seems to be a fairly obvious question: What were the factors which caused the appearance in Scotland during the second half of the nineteenth century of a number of physicists who appeared to hold shared metaphysical views and who were also engaged in the ongoing professionalisation of science? The answers to this question are not as clear as they might at first appear.

The main subjects of this discussion were all trained primarily in the physical sciences and had been exposed to significant 'Scottish' influences during their upbringing. Of course, it should be noted that the question of how Scottish identity was constructed at this period is itself somewhat problematic. Although the concept of national consciousness had generally grown in importance in western Europe following the French Revolution, Scotland at that time was not a nation state in its own right, although it did, during the first half of the nineteenth century, experience an evocation of a romanticised Highland past, as evidenced in the writings of Sir Walter Scott (1771–1832). This period also saw the proliferation of newly invented tribal tartans, which provided a sense of shared cultural affect for lowland and expatriate Scots.[10] 'Scottishness' for our subjects may in fact have been about the adoption of an identity which was in part defined by its broad opposition to the English metropolitan other, rather than being a simple adherence to a not yet fully formed national identity.

Another key influence on these individuals, at least from the points of view of several historians, is that the spaces which they inhabited in childhood and adolescence frequently brought them into intimate contact with the particular strain of Calvinism known as Presbyterianism. The idea that similarities in their cultural backgrounds, derived from shared language, religion and avocation were influential in the formation of their

[9] Sydney Checkland and Olive Checkland, *Industry and Ethos: Scotland 1832–1914* (London: Edward Arnold, 1984), 150.

[10] Hugh Trevor-Roper, 'The Invention of Tradition: The Highland Tradition of Scotland', in Eric Hobsbawm and Terence Ranger, (eds.), *The Invention of Tradition* (Cambridge University Press, 2012), 15–42.

intellectual views, has led, as noted by Bernard Lightman, to the creation of their collective soubriquet:

> Thanks to the scholarly efforts of P. M. Heimann, Paul Theerman, Norton Wise, David Wilson, and Crosbie Smith, we now know a great deal about Victorian physicists whose science was structured by religious ideas. Bearing the impress of Scottish presbyterianism and linked to the industrialists of northern Britain, energy physics was founded by a "North British" group.[11]

The work of Crosbie Smith, in particular, has been invaluable in identifying the theological drivers which underlay much of the scientific work of Thomson and his colleagues. This frequent grouping together, based on the religious and ideological similarities of several of the subjects, under such catch-all descriptions as "the more socially conservative and Presbyterian North British physicists", needs to be approached with some care, however.[12] We shall in due course unpick some of this apparent homogeneity. The North British group were more theologically diverse and less religiously enthusiastic than some commentators have assumed. Thomson's approach to religion was broadly ecumenical, whilst Jenkin remained relatively open-minded until he was close to death. In fact, other than the conversion event experienced by Maxwell (to be discussed in Chap. 10), none of the other four scientists in question exhibited significant overt religiosity in their public selves. The assumption of a strong affinity to Calvinism as an explanans for their particular physical theories should therefore not be overstated.

Scottish Education

An arguably more important shared element in their backgrounds was their exposure to the Scottish education system. It is generally acknowledged that education was highly valued and consequently of relatively high quality in Scotland from the time of the Reformation onwards, initially under the aegis of John Knox (1513–1572), the founder of

[11] Bernard Lightman, 'Victorian Sciences and Religions: Discordant Harmonies', *Osiris*, 2nd Series, 16, Science in Theistic Contexts: Cognitive Dimensions (2001): 343–366 at 352.

[12] Daniel Brown, *The Poetry of Victorian Scientists: Style, Science and Nonsense* (Cambridge University Press, 2013), 104.

Presbyterianism.[13] Although of the five people under review, two (Thomson and Jenkin) were not born in Scotland, all of them spent a significant portion of their formative years in Scottish educational institutions.

At first sight it appears that the teaching provided by a private school founded in 1824, the Edinburgh Academy, might have been the source of a shared interest in the physical sciences for several of our subjects as Martin Goldman has observed:

> The Cumming Class of the following year was even more remarkable. From it came P. G. Tait, a friend of Maxwell and Professor of Natural Philosophy at Edinburgh and Fleeming Jenkin later to be Professor of Engineering at Edinburgh,—he collaborated closely with William Thomson (Lord Kelvin) in the profitable venture of laying telegraphic cables, and also worked closely with Maxwell in standardising the measurements of electrical quantities—and his forceful criticism of the Theory of Evolution drove Darwin into Lamarckianism.[14]

Despite this interesting connection, however, it is also evident that the core curriculum of the Academy revolved mainly around classics and that the physical sciences were taught poorly, as reported by Maxwell's class-mate and future biographer Lewis Campbell (1830–1908):

> it was thought desirable that we should have lessons in "Physical Science." So one of the classical masters gave them out of a text-book. The sixth and seventh classes were taught together; and the only thing I distinctly remember about these hours is that Maxwell and P. G. Tait seemed to know much more about the subject than our teacher did.

In the same biography we find that the interest in scientific subjects exhibited by Maxwell and his peers was primarily stimulated by their parents, informed by their own reading and their access to the other cultural and academic resources to be found in post-Enlightenment Edinburgh. As an example of the latter, Campbell reported that "earlier in the same year" as the disappointing physics lessons, Maxwell had attended a lecture on

[13] Thomas M. Devine, *The Scottish Nation: 1700–2000* (London: Allen Lane, 1999), 390–391.

[14] Martin Goldman, *The Demon in the Aether, The Story of James Clerk Maxwell* (Edinburgh: Paul Harris, 1983), 31.

the discovery of the planet Neptune which "was also connected with his subsequent labours, and must have impressed him not a little at the time". The lecturer in question was John Pringle Nichol (1804–1859), who had on a previous occasion encouraged Maxwell's interest in optics.[15]

Thomson, unlike other members of the North British group, did not experience an early exposure to the intellectually stimulating influence of the Edinburgh environment. He did, however, have the alternative advantage of having been initially taught by his father, James Thomson (1786–1849), an academic who in due course was "elected to the Chair in Mathematics" at the University of Glasgow, and also by Nichol.[16] Thomson's subsequent education took place at the University of Cambridge, an institution which provides an additional, but partial, bridge between several of our actors as Maxwell and Tait both spent part of their careers there. A similar claim can be made for the University of Edinburgh, which acted as a conduit to Cambridge for Maxwell and Tait and as an employer for Stewart prior to his lengthy stint as director of the Kew Observatory. Cambridge also provides an intellectual link back to another of our earlier subjects in that it was William Whewell (1794–1866), who, amongst others, was responsible for modernising its mathematics curriculum during the 1820s, thus setting the academic agenda for some of the future North Britons.[17]

An interesting dichotomy has been suggested by Sydney and Olive Checkland: that amongst the Scottish universities in the early nineteenth century, Edinburgh "had a strong Cartesian character, running in terms of pure theory", whereas Glasgow was far more focused on "linking the academic world with that of industry". As examples of this divergence, they suggest the work of William Thomson as "an applied scientist", whereas "Maxwell's work had nothing of the practical, but was deeply philosophical". Both men had "had to go to Cambridge for their pure mathematics": an environment where they would have encountered a broader range of both philosophical and religious views.[18]

[15] Lewis Campbell and William Garnett, *The Life of James Clerk Maxwell* (London: Macmillan, 1882), 84–86. Lewis Campbell made it clear in the preface that he was the author of this section of the biography.

[16] King, *Kelvin*, 4–8.

[17] Elisabeth Leedham-Green, *A Concise History of the University of Cambridge* (Cambridge University Press, 1996), 125.

[18] Checkland and Checkland, *Industry*, 146–147. Crosbie Smith notes that Maxwell took some interest in practical crafts which aided him in his subsequent experimental work: Crosbie Smith, *The Science of Energy: A Cultural History of Energy Physics in Victorian Britain* (London: Athlone, 1998), 115.

This east-west division in terms of Scottish educational institutions should be borne in mind as we proceed to examine the topics on which our subjects were later to focus. Thomson and his colleagues tended to combine their theoretical work with more practical matters such as the laying of telegraph cables, whilst Stewart and, in particular, Maxwell, who was less financially dependent on his chosen profession, felt able to indulge in 'thought experiments', investigating such abstract topics as free will and the location of an eternity which was then seen to be threatened by the finite lifetime of the visible universe.[19]

In this and the following chapter we shall additionally point out that some identifiable changes in emphasis and occasionally novel directions in which the thoughts of these individuals carried them were preceded by significant life events which, it will be argued, were amongst the factors which caused them to rethink and 'refeel' their preceding positions. It is as though the opportunities thus provided to mentally reframe during enforced periods of 'down time' allowed them to pause and reflect on deeper questions of metaphysics and theology. This discussion of the potential impact of bodily events on overtly intellectual themes will provide further evidence of the interconnectedness of the various elements of the subject's *Weltmodell*.

William Thomson's Upbringing and Early Career

William Thomson had been born in Belfast and raised as a Presbyterian by his father. Throughout his life, William continued to be a regular church attender, but his exposure to both strands of Presbyterianism in Scotland, the Anglican liturgy in England and his elder brother's Unitarianism meant that he developed a latitudinarian approach to the varying strains of religion he encountered, as noted by Thompson:

> I am quite sure that he was sincerely religious: I would say he was a sincere Christian (meaning by Christianity the religion taught by Christ rather than the religion taught by the churches). I believe he looked deep into essentials, and that he regarded differences of sects as mere matters of form, and

[19] Maxwell inherited the estate at Glenlair following his father's death in 1856. Campbell and Garnett, *Life*, 247, 257. See further discussion in Chap. 10 of Maxwell and Stewart's more speculative ideas.

looked on the distinctions between Episcopalians, Presbyterians, Quakers, and Unitarians with supreme indifference.[20]

Thomson's mother had died when he was only six years old, and his father then took on the task of raising William and his siblings in a one-parent family: initially in Ballynahinch, County Down, and subsequently in Glasgow. "As a father, Professor Thomson was admirable. Although a strict disciplinarian, he never alienated the affection of his children."[21] When the family first moved to Glasgow, his father was paid a limited salary by the university (much of it went instead to his predecessor), and he had to support the family by giving additional lectures in astronomy and geography which were, incidentally, deemed suitable subjects for ladies.[22]

Conditions in their relatively inexpensive lodgings were unhealthy due to the marshy miasma which abounded in the vicinity of the old college buildings meaning that:

> The thick fogs which sometimes occurred in the wintertime in the neighbourhood of the Old College often made it necessary to have the gas burning indoors all day. This and the cold east winds in the spring made climatic conditions very unfavourable to the delicate.[23]

There are some clues that William's health had been significantly compromised during his younger days in this harsh environment. When he took up rowing at Cambridge, his sister, Anna, wrote to him expressing their father's concerns about his health:

> it has a little of the same spirit as gambling, he thinks, and also you will be very likely to get so much excited by the racing as to injure your health instead of improving it, which you know was the cause for which papa allowed you to get the boat.[24]

[20] Silvanus P. Thompson, *The Life of William Thomson: Baron Kelvin of Largs*, vol II (London: Macmillan, 1910), 1086–1088. His biographer is here quoting an unnamed family friend.

[21] Alexander Russell, *Lord Kelvin: His Life and Work* (London: T. C. & E. C. Jack, 1912), 12.

[22] Crosbie Smith and Matthew Norton Wise, *Energy and Empire: A Biographical Study of Lord Kelvin* (Cambridge University Press, 1989), 32.

[23] Russell, *Kelvin*, 18.

[24] Thompson, *Life*, vol 1, 65.

These concerns were echoed in a subsequent letter from his father on the topic of preparing to take his degree:

> I am always afraid some unknown or little heard of opponent may arise. Recollect, too, that you might be thrown back by illness, and that you ought therefore to be in advance with your preparation.[25]

Despite these evident concerns for William's health, it was in fact his father whose health would in due course prove to be the greater concern. James succumbed to his second bout of cholera and died on 12 January 1849 at just about the time that 24-year-old William was beginning to formulate his version of the second law of thermodynamics.[26]

William had established, at a relatively early age, a reputation as a competent astronomer and physicist and was appointed professor of Natural Philosophy at Glasgow in 1846. The dissertation which he had to write as part of the application process for this post was on a topic probably chosen by himself: 'Age of the Earth and its Limitation as determined from the Distribution and Movement of Heat within it.' This was of course a foreshadowing of one of the more controversial research topics with which he was to engage in the future.[27]

Following his appointment, Thomson began to take an interest in the novel concepts and conclusions originating from the work of several other physicists including Sadie Carnot (1796–1832), Julius Clausius (1822–1888), James Prescott Joule (1818–1889) and fellow Scot William John Macquorn Rankine (1820–1872). His contemplation of their results, which had arisen from studies of the efficiency of heat engines, led him to a conclusion which ultimately became his personal version of the second law of thermodynamics, although stated in a rather different form than is generally found in more recent texts: "it is impossible by means of inanimate material agency, to derive mechanical effect from any portion of matter by cooling it below the temperature of the coldest of the surrounding objects."[28]

With this rather dry statement, couched in the language of physics rather than natural theology, Thomson was announcing his markedly

[25] James Thomson to William Thomson, 9 April 1843, in Thompson, *Life*, vol 1, 68.

[26] Ibid., 209. Smith and Wise, *Energy*, 317–318.

[27] Thompson, *Life*, vol I, 161–189.

[28] William Thomson, *Mathematical and Physical Papers*, vol 1 (Cambridge University Press, 1882), 179.

changed understanding of a long-running dispute between two major influences in his earlier life: Chalmers and Nichol. Throughout the 1840s, he had endeavoured to reconcile their competing cosmologies. On the one hand, there was a long-standing concept of a dynamically stable universe which had been suggested by Laplace's assertion of the solar system as a smoothly running and potentially eternal engine. This cosmology supported the concept of collocation, an idea which had been promoted in a more theological and voluntarist form by Chalmers, whose church services Thomson had attended on several occasions. Nichol, for his part, was convinced of the progressive nature of the universe and that the development of different states of matter could only be interpreted in the light of a unidirectional arrow of time which separates the past from the future. Thomson's work on thermodynamics had helped to introduce the fundamental concept of entropy which in turn provided clear evidence for the existence of such an arrow.[29]

Geological and Biological Time

From mid-century onwards, there had been a clear existential element to be considered in thinking about the duration of time. The known, or assumed, quantity of time during which the universe would be able to survive in its present state faced a considerable challenge from the apparently inexorable logic of energy physics, echoed in the epigraph in which Alfred Lord Tennyson (1809–1892) equated the fate of "the dying sun" with the loss of his close friend Arthur Henry Hallam (1811–1833). If the cosmos has a finite lifetime, then the question of how eternity is to be reconciled with thermodynamics would need to be addressed by an appeal to some as yet undiscovered salvationary mechanism. Such a schema, of course, would also be consistent with revealed religion.

The developmental ideas of geologists, to which were added in due course a Darwinian concept of evolution, were promoted during the 1850s and 1860s by individuals who can be loosely grouped as scientific naturalists, several also being members of the X-Club. Their broad claim was that the current corpus of physical and biological science was sufficient to explain the observed state of the earth and the existence of life upon it, with no requirement for divine intervention. As several historians, including Lightman, Smith and Ruth Barton, have observed, these

[29] Smith and Wise, *Energy*, 84–99, 317–331.

assertions of authority were often made in magazines and journals which circulated amongst a popular audience who were not subject specialists in their own right.[30]

To oppose the certainty posited by these southern, metropolitan and broadly agnostic 'experts', several members of the North British group at various times and in different ways sought counter-evidence from the more conceptual and therefore contestable spheres which lay within their own competence, that is, science at both the very large scale and at the very small. The nascent science of energy carried important implications for the life stories of both the cosmos as a whole and its smallest components, the apparently indivisible atoms. Refuge was therefore sought by this alternative group of scientists in invisible forces and forms of matter which they hoped would sooner or later provide proofs of what Whewell had earlier called *Indications of the Creator*.[31]

The most evident and easily comprehensible (to a wider audience) line of enquiry about the cosmos with which several of our actors engaged was the question of its age. Both the age and the scale of the universe had long been subjects of quantitative debate, particularly since Ole Rømer's calculation of a finite speed of light in 1676 which in turn led to a number of estimates of the age of the universe.[32] In the second half of the nineteenth century, however, an increasing concern arose as a consequence of the evidence of ongoing change suggested by the gradually solidifying nebular hypothesis and the unveiling of the second law of thermodynamics: that alongside birth and growth there also must be decay and death. The universe had not been set in motion by its creator to run indefinitely into the future. Instead, like a clockwork machine or an automaton, it contained the seeds of its own demise: a figurative running down of the spring which also implied a starting point which may not have been as far in the past as some supporters of developmental theories required.

The age of the earth, in particular, became a subject of intense discussion in the 1860s. An upward trend in the planet's longevity which had been driven by the conclusions of geologists and astronomers had been

[30] See, for example, Ruth Barton, 'Scientific Authority and Scientific Controversy in *Nature*: North Britain against the X Club', in Louise Henson et al (eds.), *Culture and Science in the Nineteenth-Century Media* (Aldershot: Ashgate, 2004), 223–235 and Lightman, 'Victorian Sciences'. Both authors refer extensively to Smith, *Science*.

[31] William Whewell, *Indications of the Creator* (London: John W. Parker, 1846).

[32] Laurence Bobis and James Lequeux, 'Cassini, Rømer and the speed of light', *Journal of Astronomical History and Heritage*, 11(2), 2008: 97–105.

further encouraged by the work of naturalists who were seeking to understand the causes of the many apparent creations and destructions of living things and how these might have dovetailed into the various hypothesised catastrophic or uniformitarian formation processes suggested as explanations for the record in the ground. These drivers for longer chronologies had been aided and abetted by the needs of the emergent concept of biological evolution by natural selection formally introduced into the public sphere by the publication of Darwin's *On the Origin of Species* in 1859.[33]

This hypothesis was identified by early proponents of the 'conflict thesis' as a significant milestone in an alleged war between science and religion, even though similar ideas, in nascent or precursory forms, had in fact been in circulation for a considerable time, having been proposed by Erasmus Darwin (1731–1802) and Jean-Baptiste Lamarck (1744–1829).[34] Natural selection, however, was a relatively new explanation for the diversity of living and extinct species, and it was understood by Darwin, and his critics, to require considerable periods of time to carry out its shaping influences. In an endeavour to find sufficient depth of time to satisfy the requirements of his theory, Darwin, who had read and admired the work of Charles Lyell (1797–1875), resorted to an argument founded on the physical geography of a specific locale:

> Hence, under ordinary circumstances, I should infer that for a cliff 500 feet in height, a denudation of one inch per century for the whole length would be a sufficient allowance. At this rate, on the above data, the denudation of the Weald must have required 306,662,400 years; or say three hundred million years. But perhaps it would be safer to allow two or three inches per century, and this would reduce the number of years to one hundred and fifty or one hundred million years.[35]

It is interesting to note in passing that Darwin's less-frequently quoted third sentence already represented a recognition that his estimate might clash with the findings of other scientists, a threat which came to fruition in due course. Privately, and in the form of a footnote not found in the

[33] Charles Darwin, *On the Origin of Species by Means of Natural Selection, or the Preservation of Favoured Races in the Struggle for Life* (London: John Murray, 1859).

[34] See, for example, Paul Elliott, 'Erasmus Darwin, Herbert Spencer, and the Origins of the Evolutionary Worldview in British Provincial Scientific Culture, 1770–1850', *Isis*, 94, 1 (March 2003): 1–29.

[35] Denis Alexander, *Rebuilding the Matrix: Science and Faith in the 21st Century* (Oxford: Lion 2002), 181. Darwin, *Origin*, 286–287. See also Smith and Wise, *Energy*, 524.

London original but added to the first New York edition, he speculated that his "one inch per century" estimate was probably understated and might be subject to challenge. He did not have long to wait for counter-arguments to be posited against his more ambitious timescale as he himself acknowledged in the same footnote: "but I confess that an able and justly severe article, since published in the *Saturday Review* (24 December 1859), shows that I have been rash."[36]

Thomson's Intervention in Temporal Debates

At the beginning of this chapter, we outlined the serious injury which befell Thomson in late 1860, resulting in excruciating pain and an extended period of convalescence. "He is in a very helpless position, and cannot even raise his head, but he is very patient and bears it wonderfully."[37] Despite Margaret's positive report, in this letter to the German physicist Hermann von Helmholtz (1821–1894), of her husband's equanimity there is evidence that during this period he became suffused with morbid thinking and a tendency to anger. Writing to his sister Elizabeth King (1818–1896) in early March 1861, in response to a letter from her husband David (1806–1883), he was evidently much affected by the recent death of James Crum (1806–1861). According to William, Margaret, who had been present at the death bed, "has never been in deeper distress", and they had both experienced "an overwhelming paroxysm of feeling" at the loss of her uncle, "a lifelong friend". What is particularly notable about this letter, in the light of William's own suffering, is his accusatory tone as he admonished his sister for her husband's apparent lack of sympathy:

> I could not help feeling surprised and disappointed that you would seem to have understood so little the nature and severity of the blow we have suffered, and the distress which occasions not only to us but those so near and dear to us [...] I certainly expected it would be your first thought, when you heard of what had happened.[38]

[36] Charles Darwin, *On the Origin of Species by Means of Natural Selection, or the Preservation of Favoured Races in the Struggle for Life* (New York: D. Appleton, 1860) footnote on p. 252. See also: Charles Darwin to Asa Gray, 1 February 1860. Available online at: www.darwinproject.ac.uk/letter/DCP-LETT-2676.xml (Accessed 27 December 2020).

[37] Margaret Thomson to Hermann von Helmholtz, 11 January 1861, in Thompson, *Life*, vol 1, 413.

[38] William Thomson to Elizabeth King, 12 March 1861, Cambridge University Library, MS Add.8818/19.

Thomson's distress was not necessarily overstated but it is clear that he went too far in his criticism of David's response. Both Margaret and he subsequently wrote apologetically, to Elizabeth and her husband respectively, retreating noticeably from William's initial point of view and recognising that David was himself suffering from health problems at that time. Margaret's letter in particular included an appeal to scripture as consolation at a time when both she and her husband were suffering from pain and loss, foreshadowing William's subsequent reaction to his wife's own death: "'Why are you troubled?' He is still saying to us. Let us answer 'We will think and not be afraid.'"[39]

In September 1861, Thomson delivered by proxy (he had still not sufficiently recovered from his injury to attend in person) a paper entitled 'Physical considerations regarding the possible age of the Sun's heat' at the British Association meeting in Manchester in which he returned to a temporal theme that he had last visited seven years previously.[40] According to his niece: "It [the paper] had occupied him while laid up with his broken leg, and he was not able to be present and read it himself."[41] Two significant events, in addition to his serious injury, had occurred at about this time and contributed, in the opinion of Smith and Wise, to his choice of subject matter. First, Thomson had attended the famous (or infamous) debate on Darwinian evolution between Thomas Henry Huxley (1825–1895) and Samuel Wilberforce (1805–1873) which occurred during the British Association meeting at Oxford in June 1860. Secondly, he was visited by Helmholtz in August 1860. The latter event is significant in that it coincided broadly with the timescale of Thomson's decision, influenced by Helmholtz's contraction theory, to abandon the idea that the

[39] Margaret Thomson to Elizabeth King, 1861?, Cambridge University Library, MS Add.8818/20. The first embedded quote seems to be based on a partially remembered version of Luke 24:38 from a passage in which a post-resurrection Jesus is demonstrating to his disciples that he is still alive and that his body physically exists. William's apology to David King, dated 20 March 1861, is catalogued as MS Add.8818/21.

[40] William Thomson, 'On the Mechanical Energies of the Solar System.' *Edinburgh Royal Society Proceedings* [read 17 April 1854] vol III (1857): 241–244. 'Appendix on the Age of the Sun', added 15 August 1854, *Transactions of the Royal Society of Edinburgh*, vol XXI. (Edinburgh: Robert Grant, 1857), 63–80. Smith and Wise, *Energy*, 525.

[41] King, *Kelvin*, 100. The letter from which she quoted was sent by William Thomson to her father, David King, author of a popular book on geology, on 8 January 1862.

sun's heat was caused by infalling meteorites which he had advocated in 1854 but then quietly abandoned in his British Association talk.[42]

Thomson's paper was subsequently revised and published in *Macmillan's Magazine* in 1862. To a casual reader, it might have appeared that the main purpose of his argument was to suggest solar contraction as a new mechanism by which the sun's heat output could be maintained for a long, but not indefinite, period of time. The article's underlying purpose, however, was to challenge the extended timescales proposed by uniformitarian geologists during which the sun had presumably been a stable source of energy.[43] His opposition was motivated by a combination of doubts arising from his own findings in the field of thermodynamics and a concern to preserve the role of God as the director of creation for both animate and inanimate entities.[44] His final sentence, which had also terminated the 1861 presentation, constituted an almost-forlorn appeal to such a creative deity whose "great storehouse" he hoped might be the locus of novel sources of energy which would somehow enable a longer continuation of life on earth, and perhaps in the hereafter, than could then be foreseen:

> As for the future, we may say, with equal certainty, that the inhabitants of the cannot continue to enjoy the light and heat essential to their life, for many million years longer, unless sources now unknown to us are prepared in the great storehouse of creation.[45]

The author of the *Athenaeum* review of Thomson's British Association paper amplified the originally understated theological intent by adding a small but significant coda to that final sentence, "by Him who orders all things rightly and well", much to the annoyance of the author who felt that "the reporter" had thereby weakened his attempt to evince the proper sense of scientific objectivity which he felt was appropriate for the audience:

[42] Smith and Wise, *Energy*, 528–529. Thompson, *Life*, vol I, 409–417. William Thomson, 'Physical considerations regarding the possible age of the Sun's heat', *Report of the Thirty-First meeting of the British Association for the Advancement of Science; held at Manchester in September 1861* (London: John Murray, 1862), 27–28. This is a summary of Thomson's paper published in the 'Notices and Abstracts of Miscellaneous Communications to the Sections'.

[43] Thompson, *Life*, vol I, 535–551.

[44] Smith and Wise, *Energy*, 612, 634. King, *Kelvin*, 99–100.

[45] Thomson, 'Physical', 28. William Thomson, 'On the age of the sun's heat', *Macmillan's Magazine*, 5 (1862): 388–393 at 393.

> I may have to make an attack on my own paper as they have it. The last sentence is half my own and half an addition by the reporter "to give it a more religious tone". I had carefully considered the form in which, in a scientific paper, I could most effectively express the view that on purely physical grounds, I was led to.[46]

In the published article, which was targeted at a wider and more generalist audience than the 1861 paper, it became clearer that his attempt to limit the elapsed time between the formation of the sun and the present day was also being advanced in order to undermine the "three hundred million years" which Darwin had felt to be needed for evolutionary processes to unfold:

> What then are we to think of such geological estimates as 300,000,000 years for the "denudation of the Weald?" Whether it is more probable that the physical conditions of the sun's matter differ 1000 times more than dynamics compel us to suppose they differ from those of matter in our laboratories; or that a stormy sea, with possible channel tides of extreme violence, should encroach on a chalk cliff 1000 times more rapidly than Mr. Darwin's estimate of one inch per century?[47]

Thomson felt that he had derived from physical science a much stronger, possibly irrefutable, argument against great longevity for the solar system. In the light of the recent findings of thermodynamics, he was unable to identify a likely source of the sun's power which would account for an elapsed time since its birth of any more than a hundred million years. "It seems, therefore, on the whole most probable that the sun has not illuminated the earth for 100,000,000 years, and almost certain that he has not done so for 500,000,000 years."[48] Given this objective, however, he seemed unable to avoid giving the impression that his confidence in his own findings was fairly weak. Allowing an 'error bar' of five times, his datum point provided a great deal of room for manoeuvre and opened

[46] Thomson, William, 'Physical Considerations regarding the Possible Age of the Sun's Heat', *The Athenaeum*, 1771, 28 September 1861: 412–413 at 413. William Thomson to James David Forbes, 5 October 1861, Courtesy of the University of St Andrews Libraries and Museums, Special Collections, msdep7, 1861, letter 156.

[47] Thomson, 'age', 391–392.

[48] Thomson, 'age', 393.

up plenty of temporal space into which counter-attacks from a Darwinian perspective could be inserted by Huxley and others.[49]

Darwin's estimate of the age of the earth, based on "the denudation of the Weald", had almost immediately been subjected to critical attacks originating from geologists founded on discussions of erosion rates, and now faced a new form of critique in the form of the second of the two papers published by Thomson in *Macmillan's Magazine* in 1862. In this article, entitled *On the secular cooling of the earth*, Thomson supported his solar longevity argument by embarking on a determined attack on the, in his view, overstated values for the age of the terrestrial sphere which were currently being proposed. Thomson's key argument in this paper, based on his assumptions about the earth's original temperature and its rate of "secular cooling", was intended to limit the possible timescale in which the actions of geology, particularly the uniformitarian version promulgated by Lyell, and those of evolutionary biology were able to take place.[50]

"In the second paper he stated that for eighteen years it had pressed on his mind that those geologists who oppose all paroxysmal hypotheses [...] have overlooked the essential principles of thermodynamics."[51] That this long-held concern had been recently reignited and combined with a determination to attack Darwinian timescales was evident in two letters written during Thomson's convalescence in 1861. In the first letter, James Prescott Joule expressed his pleasure that Thomson was "disposed to expose some of the rubbish [Darwin's theory] which has been thrust on the public lately".[52] That Thomson specifically intended to attack Darwin's terrestrial chronology, and hence the idea of biological evolution, by means of the rate at which the earth was cooling is clear from letters written to the geologist John Phillips (1800–1874) and fellow physicist James David Forbes (1809–1868) in June 1861. In the letter to Forbes, he discussed at some length the results of measurements of electrical conductivity in different rock types: "I am chiefly interested in the subject just now in trying to prove the age of the world which by evidence of internal temperature

[49] Smith and Wise, *Energy*, 538.

[50] William Thomson, 'On the Secular Cooling of the Earth', *Transactions of the Royal Society of Edinburgh*, [Read 28 April 1862], (Edinburgh: Robert Grant & Son, 1864), vol XXIII, 167–169. Reprinted in William Thomson, *Mathematical and Physical Papers*, vol III, (London: C. J. Clay and Sons, 1890), 295–311. See also Thompson, *Life*, vol I, 538–539.

[51] Ibid., 538.

[52] Smith and Wise, *Energy*, 524–525.

cannot possibly be more than 100 or 200 million years (I mean of a period fit for organic life)."[53]

It is of course interesting, in the context of the lengthy discussion in Chap. 5, that Thomson's paper on the age of the earth implicitly accepted the assumptions made by French philosophers, Pierre-Simon Laplace (1749–1827) and Joseph Fourier (1768–1830), that the solar system had formed from a hot gaseous cloud in the manner which had subsequently been publicised in Britain by Nichol. The work of Fourier on the rate at which the earth had cooled based on observations of its current temperatures was seen by Thomson as being more empirically founded and more consistent with his understanding of the dynamic changes which the planet must have experienced than Lyell's neglect of the "essential principles of thermodynamics".[54]

It is clear that a significant portion of Thomson's ire was reserved for the ascientific and doubtful, in his view, conclusions and timescales arrived at by uniformitarian geologists and evolutionary biologists who he felt were pursuing theses based more on subjective opinions than on the basis of the kind of empirical evidence which he had adduced in support of his own argument.[55] But it is also evident that his objection to Darwinism was rooted in a theological distaste, very similar to that expressed earlier by Whewell, for any suggestion that life is not to be distinguished from non-life and that it may not have been the result of a special creative event as revealed by the 'good book' of Trinitarian faith.

The Law of Dissipation and Human Longevity

Although the writing of these two essays was motivated in part by theological considerations, in the case of the origins of the sun's energy Thomson was working with the best-available hypotheses concerning the possible sources of solar fuel. He endeavoured to deal objectively with the

[53] William Thomson to John Phillips, 7 June 1861. Quoted in Smith and Wise, *Energy*, 552. William Thomson to James David Forbes, 19 June 1861, Courtesy of the University of St Andrews Libraries and Museums, Special Collections, msdep7, 1861, letter 83.

[54] See Smith and Wise, *Energy*, 520–523, for a discussion of how Thomson adapted existing nebular theory to conform with his own continuum theory of gravity. He was also in agreement with Nichol about the eventual fate of the earth. Thomson's views of the relative reliability of Fourier and Lyell on thermodynamics are discussed, inter alia, on pages 566–567 of the same volume.

[55] Ibid., 566–568.

two contemporary favourites: a continuous in-falling of meteorites and the conversion of potential to kinetic energy as the sun contracts under the influence of gravity.[56] It is also of relevance to note that astronomers at this time had been somewhat taken aback by the apparent resistance to motion encountered by the short period comet Encke which led to a change of paradigm from empty interplanetary space to a cosmos permeated by a mysterious and resistive medium: "(or aether, as some call it)".[57] The worrying implication of the latter idea was that the orbits of planets and other bodies must inevitably also decay and spiral inwards, thus foreshadowing at least two possible deaths for the inhabitants of the earth: a cold death caused by eventual total loss of the sun's energy or a fatal contraction of the earth's orbit leading it into the fiery embrace of the sun.

As Gillian Beer and others have shown, this new-found limitation of terrestrial longevity had a significant effect on Victorian mentalities, already conditioned to a greater extent than is now normal (at least for the global north) to an anticipation of mortality.[58] Thomson had particular cause to have shared these thoughts and their related concerns with the survival of the soul: at the time when he wrote his two 'chronology' papers, not only was he recovering from his serious injury but his wife was also severely ill.

In the summer of 1852, Thomson had married Margaret whom he had known since childhood. She was: "Well educated, well read, possessed of a lively imagination and a poetic fancy, she was also of a deeply religious nature and of great earnestness of purpose." It soon became evident that at the relatively youthful age of 22 Margaret's constitution was already somewhat compromised. Earlier comments made by his sister Anna in a reply to a letter from William and later by his aunt provide tentative hints that some concerns about the state of Margaret's health may have begun to surface by 1843, at about the same time as Cordelia Whewell (1803–1855) had initially fallen ill:

> I am glad you saw Margaret Crum. Does she not look quite well that you say you hope she is not working too hard?

[56] Ibid., 526–533.

[57] Smith and Wise, *Energy*, 91–92. Thompson, *Life*, vol 1, 440.

[58] Gillian Beer, '"The Death of the Sun": Victorian Solar Physics and Solar Myth.', in J. Barrie Bullen (ed.), *The Sun is God: Painting, Literature and Mythology in the Nineteenth Century* (Oxford: Clarendon Press, 1989), 159–180.

> On July 4th Aunt Gall writes to tell William that the family has gone down to Gourock for summer quarters, and adds, good soul, "I hear Margaret Crum is home and quite well, and looking well." His fair cousin's health seems to have been not without interest.[59]

It is possible that the symptoms of tuberculosis accounted for both the frailty and the 'fairness' of William's new bride, although, as noted in Chap. 1, the ways in which nineteenth-century women's illnesses are to be retrospectively read and understood is a task which needs to be approached with caution. In any event, the fatiguing experience of an "extended tour around the Mediterranean" following the marriage proved to be the beginning of a lengthy period of illness and an ultimately childless marriage which seems to echo that of Cordelia, although there is also just a touch of cynicism in his niece's commentary: "she never recovered, and all the rest of her life was a great sufferer."[60] This comment also seems to imply that there was an element of gendered performance in the ways in which Margaret and William approached the symptoms and effects of her disease. Thomson's correspondence, for example, contained occasional reports of her intermittent improvements and setbacks accompanied by evidence that he enacted the role of a resigned but caring husband whilst taking close control of her body and her movements:

> Margaret has not advanced as I had hoped and expected she wd. have done by this time, and it has been very disappointing to feel so little improvement with so much of the summer passed. She is, however, doing somewhat better now, and I think has really advanced during the last three weeks, certainly has suffered less. She looks much better and in some respects is much better than when you saw her last, but she has not at all advanced in walking power. I always carry her up and down stairs, and often from one room to another.[61]

In 1855, William was advised to take Margaret on what turned out to be the first of several visits to the spa town of Kreuznach in Germany, the location for therapeutic treatment which had occasionally hosted the Whewells from 1851 to 1853. Thomson's report of this visit implies that

[59] Thompson, *Life*, vol 1, 66, 81, 232. See Chap. 4 for details of Cordelia's illness.
[60] King, *Kelvin*, 32.
[61] William Thomson to Elizabeth King, August 1854, Cambridge University Library, MS Add.8818/9.

Margaret's medical advisors were of a similar opinion as to the need to reduce her physical movements:

> As she is not allowed to ride, or to walk more than a few minutes every day, I take her out in a Bath-chair, between which, and a balcony on which she sits a great deal reading or working, she gets a good deal of the open air.[62]

William Whewell, as discussed at some length in Chap. 4, had experienced the similarly lengthy, intermittent and ultimately fatal illness of his wife. In the case of the Thomsons, Margaret's ill-health continued for 18 years until her death in 1870. Towards the end, her condition seems to have undergone a marked deterioration such that, in an echo of Thomson's treatment during his earlier injury and a further illustration of the limitations of the then available analgesics, her pain could not be controlled by chloroform and could barely be managed by the consumption of Nepenthe (a trade product containing opiates):

> the Chloroform did not succeed, as it produced great sickness; and scarcely even gave the freedom from pain on the second trial that it did the first. But Nepenthe has succeeded somewhat and she is having it every night, with decidedly good effect. It is the first opiate that has succeeded at all with her, and it does not seem to produce sickness and general disturbance so much as laudanum or morphia.[63]

Margaret died in June 1870. In a letter written just two months later to Thomson's sister, whose daughter had just died, he admitted that even a sense of religious consolation is somehow insufficient in the face of the immediate affect of his sense of loss:

> Yet I know too well how insufficient even the surest conviction that the loved one is happy, and the firmest hope of a transcendently happy meeting to come, are to mitigate the bitterness of the present grief.[64]

[62] William Thomson to Elizabeth King, 23 July 1855, Cambridge University Library, MS Add.8818/14.

[63] William Thomson to David King (his brother-in-law), 8 April 1870, Cambridge University Library, MS Add.8818/38.

[64] William Thomson to Elizabeth King, 18 August 1870, Cambridge University Library, MS Add.8818/42.

William Thomson, Consciousness and Atomic Theory

In the light of the concerns for his wife's health, his own recent intimation of mortality and the distress reportedly occasioned by the death of James, it is easy to understand that from early 1861 onwards Thomson actively sought for ideas and evidence that would establish the uniqueness of human life and the independent existence of non-physical components which might be construed as consciousness. The whole question of how life differs from non-life was a long-running concern for Thomson, who was keen to establish mechanisms which are unique to living things and therefore held some prospect of eternal existence outside the physical body. He had tentatively suggested the concept of a 'directing power' which consumed little or no energy in his lectures during 1852–1853, as an immaterial solution to the question of how conscious actions originate. In 1862, after he had begun his chronological assault on the geologists and evolutionists, his thoughts on this topic were further stimulated by a letter from his brother James (1822–1892), who sought a possible method by which "the mysterious influence of spirit, life or the vital principle" might be exercised whilst not being subject to the normal rules of thermodynamics.[65] We shall see in Chap. 10 how Thomson's fellow North Britons sought to come to his aid in the further development of theologically motivated arguments for both the relative brevity of the cosmos and the non-materialist underpinning of the processes of life.

A specific event which precipitated an array of responses from Thomson and his colleagues on the topic of the mechanisms which underlie the functioning of living things was the 1874 *Belfast Address* in which John Tyndall (1820–1893) asserted, amongst several other claims, that psychical phenomena were just as subject to the physical laws of causes and effect and the constraints provided by the laws of thermodynamics as were the mechanistic processes of inanimate matter. Although this was a view which had been held for some time by members of the X-Club and other scientific naturalists, it was, of course, not an opinion shared by most of their more northerly peers.

One interesting side effect of Tyndall's *Address* was the impetus it gave to a relatively recent revival of interest in Lucretius as the standard bearer of the Epicurean philosophy of atomism. This revival gained significant

[65] James Thomson to William Thomson, 7 April 1862, quoted in Smith and Wise, *Energy*, 619.

impetus following the appearance in 1864 of an English translation of '*De Rerum Natura*' by Hugh Andrew Johnstone Munro (1819–1885), followed by two significant contributions published in 1868: Tennyson's poem '*Lucretius*' and a review of Munro's work by Jenkin.[66]

Following Tyndall's intervention, a number of "expressly Christian commentaries" by "John Veitch, John Masson, W. H. Mallock and others" engaged in dialectical processes of resurrecting and then refuting the views of the Roman philosopher on the grounds of scientific inaccuracy.[67] In many ways, this was an exercise in futility, the intellectual equivalent of a Quixotean tilting at windmills, since it is apparent from a reasonably close reading of Tyndall's speech that Lucretius was not being engaged as the ultimate authority in the cause of atomism but was being employed as just such a Hegelian literary device. His role was to provide a mouthpiece for a particular thesis, to which Bishop Butler was the antithesis, from which Tyndall drew a synthesis in support of his broadly pantheistic conclusions.[68]

Nonetheless, renewed interest in Lucretius' version of atomic theory coincided neatly with a parallel debate amongst scientists about the true nature of atoms. This debate had extensive roots in another ongoing disagreement between protagonists who posited a universe composed entirely of matter and those who maintained, with Thomson, that there was still room, or even a requirement, for the existence of immaterial entities such as mind or spirit. Over time Thomson had developed a view which held that atoms are in fact miniature whirlpools originating within the "perfect fluid all through space", that is, the ether. In 1867, he asserted that such fundamental building blocks of matter are "indestructible" and that therefore "they had the additional advantage that only God could create them". This idea, which had been supported by his recent correspondence with Helmholtz and subsequent experiments with smoke rings, was further supported by Jenkin's lengthy essay on Lucretius.[69]

[66]Frank M. Turner, 'Lucretius among the Victorians', *Victorian Studies*, 16, 3 (March 1973): 329–348 at 334–335. Fleeming Jenkin, 'Lucretius and the atomic theory' in Sidney Colvin and J. Alfred Ewing (eds.), *Papers, Literary, Scientific etc. by the late Fleeming Jenkin, F.R.S., LL.D.* vol 1 (London: Longmans, Green, 1887), 177–214.

[67]John Holmes, 'Lucretius at the Fin de Siècle: Science, Religion and Poetry', *English Literature in Transition*, 1880–1920, 51, 3, (2008): 266–280 at 267. Turner, 'Lucretius', 340–348.

[68]Turner, 'Lucretius', 336–337. Ruth Barton, "John Tyndall, Pantheist: A Rereading of the Belfast Address', *Osiris*, 2nd Series, 3 (1987): 111–134.

[69]Smith and Wise, *Energy*, 417–419. Jenkin, 'Lucretius', 185–188.

Thomson's hypothesis suggested as a corollary that if atoms are not interconvertible, but are in fact indestructible, then they must all have been created by some transcendent power and the total stock could not now be increased or decreased by human effort; a conviction which helps to explain his subsequent rejection of the idea that radioactivity could be the source of the sun's energy.[70] His motivation for continuing to support the 'vortex atoms' theory against the possibility of radioactive decay during later life was also likely to have included a distaste for having much of his earlier work on this topic deprecated.[71]

Another strong supporter of the idea of indestructible created matter was a fellow North Briton, Maxwell, who claimed no less an authority than that of John Frederick William Herschel (1792–1871) for his assertion that molecules had "the exact quality of a manufactured article" rather than having in some way evolved from precursory ingredients as had been claimed by Herbert Spencer (1820–1903), Tyndall and others.[72] He observed that the universal nature of matter had been recently demonstrated by the discovery of the spectra of "stars so distant that no material thing can ever have passed from one to another":

> Each molecule, therefore, throughout the universe, bears on it the stamp of a metric system as distinctly as does the metre of that Archives at Paris, or the double royal meter of the Temple of Karnac.[73]

This evidence appears to have been partly adduced in support of Maxwell's disagreement with the claims made by Charles Piazzi-Smyth (1819–1900) of ancient British connections with the metrology of the Great Pyramid, but it was clearly also a building brick in his vision of a cosmos in which there is "a *collocation*, to use the expression of Dr. Chalmers" of matter whose dimensions, locations and motion had been determined by an omnipotent architect: "Him Who in the beginning created not only the heaven and earth, but the materials of which heaven and earth consist".[74]

[70] Smith and Wise, *Energy*, 549–551, 607–611.

[71] Thompson, *Life*, vol 1, 510–519. Thompson, *Life*, vol 2, 1190.

[72] Smith and Wise, *Energy*, 535, 645. Matthew Stanley, 'The Pointsman: Maxwell's Demon, Victorian Free Will, and The Boundaries of Science', *Journal of the History of Ideas*, 69, 3, July 2008: 467–491, at 476–477.

[73] James C. Maxwell, 'Molecules', *Nature*, 8, 25 September 1873: 437–441 at 441.

[74] Simon Schaffer, 'Celestial Eccentric', *Nature*, 333, 2 June 1988: 405–406. Eric M. Reisenauer, '"The battle of the standards": Great Pyramid metrology and British identity', 1850–1890, *The Historian*, 65 (2003): 931–978. Maxwell, 'Molecules'.

Chronology, Chloroform and Calvinism

William Thomson wrote the first in a series of papers which argued for limited longevity of the sun and the earth in furtherance of at a time when he was recovering from a severe and very painful fracture. Thus, it would be reasonable to posit a conjunction of causes, including the significant injury and continuing concerns for his wife's well-being as well as his personal religious views and his distaste for lengthy geological and biological chronologies, which actuated his renewed focus on some of his earlier thoughts about solar and terrestrial longevity.

There is an additional potential causal factor which may have motivated his turn towards chronology debates. Having broken his leg in the fall described earlier, he was exposed to "very large doses" of chloroform which were intended to dull the pain:

> He never showed any tendency to fever, but suffered much from the effects of the chloroform, of which it required very large doses to make him insensible. He suffers much uneasiness and considerable pain. The pain caused by setting the limb was very great.[75]

The use of chloroform as an anaesthetic was a relatively recent development, and, as we saw in the case of Margaret Thomson, for some patients the side effects and limited effectiveness restricted its utility. Its potential effectiveness had been established in 1847 by the Scottish obstetrician James Young Simpson (1811–1870), who initially exposed himself and several guests to it as a post-dinner entertainment. One young lady who volunteered to try it, "a Miss Petrie", subsequently "fell asleep crying 'I'm an angel. Oh, I'm an angel'!"[76] The reported effects of chloroform varied from subject to subject, depending on dosage and context, in a manner similar to initial experiences of opium consumption. For Herbert Spencer, its use during a dental procedure seems to have resulted in the shutting down of virtually all forms of consciousness without freeing him from the sensation of pain. Cognitive processes under these circumstances were

[75] Margaret Thomson to Hermann von Helmholtz, 11 January 1861, in Thompson, *Life*, vol 1, 413.

[76] H. Laing Gordon, *Sir James Young Simpson and Chloroform (1811–1870)*, (London: T. Fisher Unwin, 1897), 106–108.

distinctly not 'angelic'.[77] Other patients, as documented by Mike Jay, reported less uncomfortable but somewhat disorienting "out-of-body" experiences:

> One doctor had experienced an "exactly similar sensation" himself; another had been told by a patient that "I thought I had got to the bottom of the secrets of nature". Another recalled a further version of the Perennial Philosophy of the inhalant user; "When under chloroform, the Platonic idea came to me that Matter was only phenomenal, while the only reality was that which underlay Matter—viz. its spiritual substance".[78]

The reported effects of chloroform were sufficiently well known towards the end of the century that the author Herbert George Wells (1866–1846) employed them in a fictional description of a similar excursion which led the narrator into a "vast cosmic perspective":

> his successful short story *Under the Knife* begins with the protagonist inhaling chloroform on the operating table before being precipitated into an out-of-body voyage above the earth, beyond the stars and finally to a vantage point where the entire galaxy is revealed as a mote of dust in the disembodied hand of a mysterious Creator.[79]

In the light of these reported experiences, it is quite possible that Thomson's substantial doses of chloroform may similarly have produced disorienting experiences and altered states of mind freighted with religious and metaphysical affect.

Thomson had also been treated with opiates during the earliest and most painful period of his injury. As has already been noted in Chap. 1 the users of extracts from *Papaver somniferum*, especially during the earlier phases of consumption, generally experience vivid 'opium dreams' during which they are exposed to dramatically altered perceptions of time and space. This linkage is of some significance when seeking the root causes of Thomson's renewed focus on questions of cosmic chronology which

[77] Herbert Spencer, 'Consciousness under chloroform', *Popular Science Monthly*, 13, October 1878: 694–698.

[78] Mike Jay, *Emperors of Dreams: Drugs in the Nineteenth Century* (Sawtry: Dedalus, 2000), 145.

[79] Ibid., 145. The short story in question was originally published in 1897 and reprinted as: Herbert G. Wells, 'Under the Knife', *Tales of the Unexpected* (London: W. Collins Sons, 1924), 54–70.

became evident during his convalescence. As with the other interactions between injury and ideology documented in this book, there is an arguable case for a connection between Thomson's lengthy recuperation from a very painful injury, which afforded him an ample opportunity for reflection, and his subsequent focus on a topic which allowed him to bring his fire to bear on what he regarded as the theologically negative consequences of Darwin's evolutionary theory

It is also quite possible that the ecumenicism displayed in his public proclamations on the topic of religion, and which do not bespeak a strong confessional alignment, had not completely overwritten the religious influences he experienced during his early upbringing in Belfast and then Glasgow. His exposure to Evangelical Presbyterianism may have left a deeper imprint than his subsequent comments might suggest. It is of interest, for example, to note that the local church which Thomson attended as a boy in Ireland had been funded by his father and inaugurated by a well-known Evangelist: "The famous Dr. Chalmers was invited to come from Scotland to open the church, and he stayed about a week with us on the occasion."[80] A reasonable response to the events of 1860–1862 might be to infer that the outwardly rational latitudinarian self which was an evident component of his adult *Weltmodell* was forced to contend at times of great stress with an earlier and more evidently Calvinist construct which surfaced from time to time and influenced the selection of topics of interest to his scientific self.

[80] Elizabeth T. King, *Lord Kelvin's Early Home* (London: Macmillan, 1909), 47.

CHAPTER 10

The North Britons: Physics and Metaphysics

> And to add a new weirdness to what the sky possesses in its size and formlessness, there is involved the quality of decay. For all the wonder of these everlasting stars, eternal spheres, and what not, they are not everlasting, they are not eternal; they burn out like candles.
> Thomas Hardy, *Two on a Tower*[1]

We saw in the previous chapter that, following a serious injury in 1860, William Thomson (1824–1907, later Lord Kelvin) began a theologically tinged 'campaign', derived primarily from consideration of the laws of thermodynamics, to limit the possible ages of the earth and the sun which seemed to be implied by the work of geologists and required by the evolutionary theory of Charles Darwin (1809–1882). One of William Thomson's most vocal supporters was Henry Charles Fleeming Jenkin (1833–1885), the Regius Professor of Engineering at the University of Edinburgh. Jenkin had been born in Kent but was educated in Scotland, firstly at Jedburgh School and then at the Edinburgh Academy. At the latter seat of learning, he had encountered two other future participants in thermodynamic and temporal debates: James Clerk Maxwell (1831–1879) and Peter Guthrie Tait (1831–1901).[2]

[1] Thomas Hardy, *Two on a Tower* (London: Macmillan, 1923), 66.

[2] Robert L. Stevenson, 'Memoir' in Sidney Colvin and J. Alfred Ewing (eds.), *Papers, Literary, Scientific etc. by the late Fleeming Jenkin, F.R.S., LL.D.* vol 1 (London: Longmans, Green, 1887), xi–cliv, xxix.

H. Carlton, *Cosmology and the Scientific Self in the Nineteenth Century*, https://doi.org/10.1007/978-3-031-05280-4_10

Although brought up originally in the Church of Scotland, Jenkin seems to have preferred a relatively open-minded approach to religion, evincing a "fine spirit of piety" but eschewing "dogma". His slightly sceptical but nonetheless receptive stance is perhaps best captured in a single sentence from a letter written to his friend Robert Louis Stevenson (1850–1894):

> I cannot conceive that any single proposition whatever in religion is true in the scientific sense; and yet all the while I think the religious view of the world is the most true view.[3]

This statement, which neatly encapsulates Jenkin's understanding of the epistemic relationship between science and religion, would appear to place him in a slightly different metaphysical frame than some of his collaborators in the group of scientists who have subsequently been labelled the 'North British physicists'.

Stevenson was significantly younger than Jenkin whom he had first encountered as a student when he found him to be more personally engaging and rather more approachable than other academics:

> His presence was against him as a professor: no one, least of all students, would have been moved to respect him at first sight: rather short in stature, markedly plain, boyishly young in manner, cocking his head like a terrier with every mark of the most engaging vivacity and readiness to be pleased, full of words, full of paradox, a stranger could scarcely fail to look at him twice, a man thrown with him in a train could scarcely fail to be engaged by him in talk, but a student would never regard him as academical.[4]

As a self-defined "youthful atheist", he found the older man's more complex views on both science and religion intellectually challenging and refreshing: "the atheistic youth was met at every turn by a scepticism more complete than his own".[5] We are told that over time, however, Jenkin's "conformity to the church in which he was born grew more complete, and his views drew nearer the conventional". Stevenson went on to describe Jenkin's broad *rapprochement* with the beliefs of the Church of Scotland towards the end of his life, a perhaps not unexpected development:

[3] Stevenson, 'Memoir', cxxxiv–cxxxvii.
[4] Stevenson, 'Memoir', cxxxii.
[5] Ibid., cxxxv–cxxxvi.

'The longer I live, my dear Louis,' he wrote but a few months before his death, 'the more convinced I become of a direct care by God—which is reasonably impossible—but there it is.' And in his last year he took the communion.[6]

Stevenson's memoir contains a detailed account of Jenkin's earlier years which turn out to have been unusually variegated, including an extended 'exile', brought about by his father's fall into comparative poverty. During a sojourn in France, which lasted until early 1848, his family "whose politics were Liberal" witnessed some of the events of the 'February Revolution'. Jenkin was an initially sympathetic bystander who observed and commented on the ongoing upheaval. He provided light-hearted, almost insouciant, descriptions of the events in letters written "to influence his friend Frank, an incipient Tory". He continued to be present on the streets as the barricades went up, and he experienced at first hand the protest marches, the violence of state suppression and the subsequent fighting which led to "the insurrection":

The Revolution, [...] was quite delightful: getting popped at and run at by horses, and giving sous [coins] for the wounded into little boxes guarded by the raggedest, picturesquest, delightfullest, sentinels; but the insurrection! ugh, I shudder to think at it.

The increasingly volatile situation brought about by the second wave of uprising which was of a more proletarian nature than the preceding, primarily bourgeois, 'revolution' was judged to be too dangerous by his parents. They escaped from Paris and made their way to Genoa where they experienced further episodes of popular uprising.[7] It is interesting to note, particularly in the light of some of the dramatic theological conversion narratives which affected some of the subjects of this book, that Jenkin's early exposure to such violent contemporary instances of the suppression of revolutionary movements did not seem to have brought about any significant leftward shift in his political thinking.

His youthful experiences did, however, appear to act as a spur to his later engagement with some slightly unexpected topics, including trade union law and economic theory, as he developed an unusually wide range

[6] Ibid., cxxxv.
[7] Stevenson, 'Memoir', xxxi–xlv.

of interests when compared with his Scottish peers. His five papers on political economics, written from a Liberal perspective, began with 'Trade Unions: how far legitimate', published in 1868. This was followed by further work which included the introduction of graphical methods into economics and the first formulation of modern wage theory. His significant contributions to the discipline were subsequently recognised by Joseph Schumpeter (1883–1950), who asserted that Jenkin was "an economist of major importance, whose main papers [...] form an obvious stepping stone between J. S. Mill and [Alfred] Marshall".[8]

Fleeming Jenkin and the Chronology Debate

From the point of view of our main theme, it is potentially significant that two years prior to his first intervention in the chronology and evolution debates, and their associated religious subtext, Jenkin suffered a serious setback. According to Stevenson:

> In September of the next year [1865], with the birth of the second [child], Charles Frewen, there befell Fleeming a terrible alarm and what proved to be a lifelong misfortune. Mrs. Jenkin was taken suddenly and alarmingly ill; Fleeming ran a matter of two miles to fetch the doctor, and drenched with sweat as he was, returned with him at once in an open gig.

As a consequence of this event and a night spent "crouching on the floor in the draught, and not daring to move lest he should wake the sleeper", Jenkin himself succumbed to a combination of disabling conditions. It is, however, unclear as to what extent his subsequent difficulties were the result of the events of that single night or had been nascent in an existing lack of "vigour" but Stevenson clearly saw a causal connection, leading to Jenkin's subsequent evocation of stoic resignation:

> He had never been strong; energy had stood him instead of vigour; and the result of that night's exposure was flying rheumatism varied by settled sciatica. Sometimes it quite disabled him, sometimes it was less acute; but he was rarely free from it until his death, I knew him for many years [...] and

[8] Joseph A. Schumpeter, *History of Economic Analysis* (New York: Oxford University Press, 1954), 837 n10. All five papers can be found in Sidney Colvin and J. Alfred Ewing (eds.), *Papers, Literary, Scientific etc. by the late Fleeming Jenkin, F.R.S, LL.D.*, vol 2 (London: Longmans, Green, 1887), 3–154.

> during all this time, he only once referred to his infirmity and then perforce, as an excuse for some trouble he put me to, and so slightly worded that I paid no heed. This is a good measure of his courage under sufferings of which none but the untried will think lightly.[9]

This resigned acceptance of earthly difficulty, which Stevenson attributed to pious acquiescence, "old-fashioned morality" and "simple faith", seems to fit the patterns for the performance of suffering outlined in Chap. 1 and is a faint echo of the experiences of some of our other subjects, in particular the pious and passive approach to death of Cordelia Whewell (1803–1855, née Marshall).[10] The attitude of resignation which Stevenson reported also fits quite well, especially when combined with the approach of death, with his gradual adoption of a more conventional religious self, culminating in his eventual resolution with the faith of his childhood.

There are no contemporary reports which suggest that there was any immediate change in Jenkin's theological, scientific or metaphysical stances following this nocturnal misadventure despite its evident impact on his future health. It was clearly not a life-threatening event (at least not from Jenkin's point of view, although he may well have feared for the lives of his wife and child). The onset of rheumatism and sciatica, which from time to time "quite disabled him", did not represent an existential threat but does, incidentally, parallel the earlier experience of William Thomson who was had been permanently disabled by an accident in late 1860. It is interesting to note that it was only after the onset their respective disabilities that Thomson and Jenkin chose to weigh in on the subject of evolutionary timescales and their claimed improbability. Jenkin's 1867 review of Darwin's *Origin of Species* in the *North British Review* strongly supported the position adopted earlier by his long-standing colleague and mentor, who had initiated this debate in 1861, so it has to be said that Jenkin was a little late to the fray.

Jenkin had been collaborating with Thomson from 1859 onwards on the submarine cable projects with which the more experienced physicist had already been engaged for several years. Their joint activities subsequently extended to several other spheres, including his "partnership with Sir William Thomson and Mr. Varley in many ingenious patents" which

[9] Stevenson, 'Memoir', lxxii–lxxiii.

[10] Ibid., lxiii. See Chap. 1 on Victorian experiences of death and Chap. 4 for the story of Cordelia's terminal illness.

fortunately began to bear fruit in 1868 as he had for some time been labouring under "clouds" of "pecuniary embarrassments" arising from his ill-fated partnership with "Mr. Forde, a gentleman in a good way of business" which had not proven to be particularly fruitful. His personal circumstances were further enhanced a little later when he "was appointed to the new chair of engineering at the University of Edinburgh."[11] It was therefore at a time when Jenkin was still experiencing financial difficulties that he first chose to support Thomson's thermodynamical assault on long chronologies.

In the 1867 article, Jenkin assembled several arguments which he felt rendered Darwin's hypothesis of development unlikely or improbable, including Thomson's demonstrations that the ages of both the sun and earth are far from infinite and most likely too short to permit Darwin's processes of natural selection to work through as claimed. He built what he felt to be a solid and reflexive case in which he assumed that posteriority would "look back with [...] admiration" on Darwin's "ingenious and plausible speculation" in a fashion similar to the bemused tolerance extended by his contemporaries to "the atoms of Lucretius, or the crystal spheres of Eudoxus":

> A plausible theory should not be accepted while unproven; and if the arguments of this essay be admitted, Darwin's theory of the origin of species is not only without sufficient support from evidence, but is proved false by accumulative proof.[12]

Although Jenkin demurred from identifying a specific duration during which life on earth had been feasible from an energetic perspective, he nonetheless asserted that the "age of the inhabited world is proved to have been limited to a period wholly inconsistent with Darwin's views".[13] His case for long chronologies was made with such power that Darwin himself was moved to acknowledge the influence of Jenkin's writing: "Fleeming Jenkin has given me much trouble, but has been of more real use to me than any other essay or review."[14]

[11] Stevenson, 'Memoir', lxvii, lxxiii.

[12] Colvin and Ewing (eds.), *Fleeming Jenkin*, vol 1, 263.

[13] Ibid., 248.

[14] Charles Darwin to Joseph D. Hooker, 16 January 1869, in Francis Darwin and Albert C. Seward, (eds.), *More Letters of Charles Darwin*, vol II (London: John Murray, 1903), 378–379.

Peter Guthrie Tait's Support for William Thomson

Peter Guthrie Tait (1831–1901), on the other hand, was much less reticent about declaring fixed limits for the ages of the earth and sun. He had been a classmate of Jenkin at the Edinburgh Academy but, unlike his peer, he was a convinced Trinitarian. "His reverence for the undoubted essentials of the Christian Faith was deep and unmovable; and nothing pained him so much as a flippant use of a quotation from the Gospel writings." Tait's general approval of the "rational claim on our belief" and the democratic and universalist nature of Christianity, as he understood it, can easily be detected in this excerpt from an article which he wrote in 1878:

> it would appear that the only religion which can have a rational claim on our belief must be one suited equally to the admitted necessities of the peasant and of the philosopher. And this is one specially distinguishing feature of Christianity. While almost all other religious creeds involve an outer sense for the uneducated masses and an inner sense for the more learned and therefore dominant priesthood, the system of Christianity appeals alike to the belief of all[15]

Tait thus acts as the base case or conservative hypothesis against whom the other North Britons can be compared, the 'exception which proves the rule'. Given a consistent and convinced belief in Trinitarian theology, he had no need of a *Weltmodell* crisis to motivate his inclination to support Thomson's arguments against long chronologies and scientific naturalism when the occasion arose.[16]

Tait eventually became professor of Natural Philosophy at the University of Edinburgh in which role he collaborated closely with Thomson, who held the same chair at the University of Glasgow. They jointly commenced the writing of a series of books under the umbrella title of *A Treatise on Natural Philosophy*. This collection, generally referred to as 'Thomson and Tait' or sometimes just 'T & T', was intended to be a comprehensive survey of the physical sciences. It was, unfortunately, never completed in

[15] Cargill G. Knott, *Life and Scientific Work of Peter Guthrie Tait* (Cambridge University Press, 1911), 35–37. On page 36 Knott quoted from Peter G. Tait, 'Does Humanity demand a New Revelation', *International Review* (November 1878).

[16] A German word meaning 'world picture' which has previously been employed to describe both the cosmological models of physical scientists and the creation stories found in traditional religions. It thus encompasses the two key epistemologies under discussion here.

totality: the first two volumes were published in 1867 but the planned series eventually ground to a halt as the co-authors were diverted by other projects.[17]

Tait, "the bulldog of the North British circle", had engaged in debates on the topics which divided the North British group from their metropolitan counterparts both before and after the *Belfast Address* given by John Tyndall (1820–1893) in 1874.[18] In part, he seemed to have been motivated by jealousy of Tyndall's popularity and commercial success. But it is also evident, as Crosbie Smith and Norton Wise have noted, that Tyndall had for a long time been one of the chief targets of Thomson and Tait's opposition to "so-called naturalism". According to Smith and Wise:

> By 1862, Thomson's convictions had become sufficiently developed for him to present, along with Tait, a popular account of his cosmos, directed principally against John Tyndall (1820–1893), in the Presbyterian magazine *Good Words*. Tyndall's so-called naturalism, which assigned no role to God, and which placed all things (including mind) under the rule of deterministic, materialistic law, regularly provoked the wrath of his Scottish contemporaries [...] as did his failure to comprehend the Second Law of Thermodynamics.[19]

In a later contribution in support of Thomson's chronological minimalism, Tait submitted a discussion of 'Geological Time', which was published in the *North British Review* (1869). This was followed seven years later by an essay on thermodynamics contained in a larger volume entitled *Some Recent Advances in Physical Sciences* (1876).[20] In both publications, he was prepared to argue for a much shorter duration of life on earth than that implied by any of Thomson's earlier estimates.

In 1869, Tait appeared to have been personally offended by the attitude of Thomas Henry Huxley (1825–1895), who had been recently

[17] Silvanus P. Thompson, *The Life of William Thomson: Baron Kelvin of Largs*, vol I (London: Macmillan, 1910), 477–479.

[18] Ruth Barton, 'Scientific Authority and Scientific Controversy in *Nature*: North Britain against the X Club', in Louise Henson et al. (eds.), *Culture and Science in the Nineteenth-Century Media* (Aldershot: Ashgate, 2004), 223–235 at 230.

[19] Crosbie Smith and Matthew Norton Wise, *Energy and Empire: A Biographical Study of Lord Kelvin* (Cambridge University Press, 1989), 535.

[20] Peter. G. Tait, 'Geological Time', *North British Review* 50 (1869): 406–439. Peter G. Tait, *Lectures on Some Recent Advances in Physical Science*, 2nd Edition (London: Macmillan, 1876), 160–175.

appointed as President of the Geological Society. In his inaugural address, as recollected by Tait, Huxley had asserted, "I am as incompetent to judge the evidence on which *you* go as you are to judge *ours.*" Tait interpreted this statement as being "equivalent to telling mathematicians and natural philosophers, [to] mind their own business, and let other folk be". From Tait's perspective, the style and content of Huxley's rejection of the northern physicists' reduced timescales was uncompromising, peremptory and irrational. Being, like several of his colleagues, "fundamentally conservative", he compared this attack with the apparent anarchy of recent industrial disputes: "those most objectionable theories and practices of the Trades-Unionists which have recently been held up to public execration". Tait's 1869 article on geological time was also intended as a riposte to what he saw as Huxley's attacks on Thomson by means of "venomous but absurd attempts at smartness".[21]

In both the 1869 paper and his 1876 book, Tait decided to tighten the window during which life could have existed to a level at which Darwin would have great difficulty in demonstrating that the process of natural selection could have operated in the way he had described. This approach was not well received by committed evolutionists. As the English physicist Joseph David Everett (1831–1904) put it in a letter to Nature:

> Prof. Tait, in his "Recent Advances in Physical Science," recapitulates these [Sir W. Thomson's] arguments, but with a different conclusion. He states the limit of age to be about *ten* million years. As the subject is of immense interest, may I ask Prof. Tait to explain this change of conclusion?[22]

Everett was here being a little disingenuous as Tait did in fact adduce three arguments, which had previously been aired in the 1869 article, to support his conclusions. Of these, the second argument is perhaps the most interesting as it involved a reimagining of time in a way which would have been familiar to a Victorian astronomer but quite novel to the average reader. Building on the earlier conclusions of Pierre-Simon Laplace (1749–1827), Immanuel Kant, (1724–1804), John Couch Adams (1819–1892) and Thomson himself, Tait observed that the rate of the earth's rotation is slowing and the moon's orbital period is increasing (technically known as the secular acceleration of the moon's mean motion)

[21] Tait, 'Geological', 406–407, 439. Knott, *Tait*, 37.

[22] Joseph D. Everett, 'Letter', *Nature*, 20 April 1876, 486.

such that, say, a billion years ago the earth would have turned on its axis in 12 rather than 24 hours. His article pointed out that this diminution of the length of a day is an indication of the relativity of chronology and the impermanence of the universe as we now see it. Since the earth's 'figure' was not as distorted as required, in Tait's view, to support a great speed of rotation he concluded that it cannot have been solid for a long period of time:

> We say then, that because the earth is so little flattened it must have been rotating at very nearly the same rate as it is now rotating, when it became solid. Therefore [...] it cannot have been many millions of years back when it became solid, [...] That argument, [...] probably reduces the possible period which can be allowed to geologists to something less than ten millions of years.[23]

The tone of Tait's comments here, in common with some of his other key points, was far from objective: he was clearly allowing a thinly disguised subtext to drive his arguments. Without showing his detailed working, he confidently postulated a timescale which he believed to be impossibly short for either uniformitarian geology or Darwin's hypothesis to be feasible, to the evident and understandable frustration of fellow physicist Everett.[24] George Howard Darwin (1845–1912), the son of Charles, was willing to accept the valuable contribution to the definition of geological time made by Thomson but argued with some vigour in his 1886 presidential address to the British Association that Tait's revamped timeline was based on uncertain estimates in which the margin of error was "far too great to justify us in accepting such a narrowing of the conclusions".[25]

James Clerk Maxwell and Free Will

In his own endeavours to undermine longer chronologies, Thomson had invoked the ideas of thermodynamics to demonstrate that both the past and the future longevity of the portion of the universe inhabited by man

[23] Thompson, *Life*, vol 1, 540–543. Tait, *Lectures*, 174.

[24] Tait, *Lectures*, 167 (and footnote) makes it clear that Tait's target in this essay was still the three hundred million years posited originally by Darwin.

[25] Smith & Wise, *Energy*, 602. The quotation comes from George Darwin's Presidential Address to the British Association for the Advancement of Science in 1886.

must, of necessity, be limited. Living things did not represent an exception to the general principle of increasing entropy, despite their ability to organise matter and display evidence of volition: "Thomson held that living creatures and dead matter were distinguished by the former's privileged possession of free will, a will which enabled them to direct the energies of nature in accordance with the dissipation principle."[26]

Thomson's elucidation of the second law of thermodynamics had, however, provided further ammunition for scientific naturalists, many of whom, in his opinion, were attempting to portray living things as purely deterministic entities subject to the same physical and chemical processes as inorganic matter and devoid of any spiritual processes which might be termed 'life' or 'free will'.[27] And if life was not a special attribute of living things but simply an emergent property of known or knowable rules identified by the sciences, then the more tenuous concept of 'consciousness' might itself be a physiological process which does not require any component which can exist independently outside the physical confines of the body or the temporal bounds of its lifespan.

Thomson was concerned by these mechanistic viewpoints but was not readily able to refute them. By adhering to the logic of his own work, he found it difficult to speculate, except in very abstract terms, about novel mechanisms which might help to identify a space in which free will or consciousness could operate. The production of ripostes to the metropolitan materialists he consciously or unconsciously delegated to those Caledonian colleagues and fellow travellers who shared his conviction that, despite recent advances in biology, physiology and an emerging physicalist science of emotions, there nonetheless remained essential differences between living and non-living entities.[28]

James Clerk Maxwell (1831–1879) made several attempts to assist Thomson by explaining the operation of free will. Maxwell was born in Edinburgh and was initially educated at the Edinburgh Academy, then at

[26] Ibid., 612.

[27] See, for example, the discussion of: Herschel to the Bishop of Kingston, 14 July 1865, Royal Society Herschel MSS HS 24.118 in Chap. 7.

[28] Thomson's thoughts on the energetics of free will are summarised, for example, in: Thompson, *Life*, vol I, 229–230 and Smith and Wise, *Energy*, 612–618. For the history of physicalist theories of emotions in this period, see: Thomas Dixon, *From Passions to Emotions: The Creation of a Secular Psychological Category* (Cambridge University Press, 2003), 135–146.

the University in the same city.[29] His 'theological career' was in some ways comparable to the direction of travel followed earlier by Thomas Chalmers. He had been exposed to both Anglican and Presbyterian liturgies as a boy, which meant that on his subsequent entry to the University of Cambridge he projected an ecumenical and rationalist self when faced with questions of religion, as demonstrated by the lightness of touch and slightly quirky sense of humour displayed in some of the essays which he wrote as a member of the 'Apostles': "an elite literary and philosophical discussion group".[30] His Broad Church approach was epitomised in a letter written in 1852 to his friend, the future cleric Lewis Campbell (1830–1908), in which he contrasted the lack of taboo topics of discussion in mainstream Christianity with the "the territory of the Scoffer, the Pantheist, the Quietist, Formalist, Dogmatist, Sensualist, and the rest":

> Christianity that is, the religion of the Bible is the only scheme or form of belief which disavows any possessions on such a tenure. Here alone all is free. You may fly to the ends of the world and find no God but the Author of Salvation. You may search the Scriptures and not find a text to stop you in your explorations.[31]

At a critical period in his academic career, at the age of 22 and just prior to his third-year examinations, however, Maxwell suffered a lengthy illness whilst he was staying at the house of the uncle of one of his fellow students, the Reverend Charles Benjamin Tayler (1797–1875), an Evangelical Anglican who was an ardent opponent of the more formulaic approach of the Tractarians.[32] As a consequence of his illness and the subsequent period of convalescence in the minister's house, Maxwell's religious views were significantly intensified. This episode had a profound effect on his future thinking on the relative epistemic and emotional weights which should be accorded to "Love" and "Knowledge". His biographers, Campbell and

[29] Lewis Campbell and William Garnett, *The Life of James Clerk Maxwell* (London: Macmillan, 1882), 2, 47, 105.

[30] Paul Theerman, 'James Clerk Maxwell and religion', *American Journal of Physics* 54, 312 (1986): 312–317 at 312. See Campbell and Garnett, *Life*, 223–246, for excerpts from some of Maxwell's essays for the Apostles.

[31] Ibid., 179.

[32] Charles B. Tayler, *Sermons for All Seasons, Chiefly on the Subject of Tractarian Error* (London: Sampson Low, 1850), vii–xiii.

William Garnett (1850–1932), reported the experience as "having given him a new perception of the Love of God":

> The illness is described by Mr. Tayler [the Reverend Tayler's nephew] as a sort of brain fever, and he was disabled by it for more than a month. The Taylers nursed him as they would have nursed a son of their own, and Maxwell, in whom the smallest kindnesses awakened lasting gratitude, was profoundly moved by this. He referred to it long afterwards as having given him a new perception of the Love of God. One of his strongest convictions thenceforward was that "Love abideth, though Knowledge vanish away." And this came to him at the very height of the intellectual struggle.[33]

The impact which the interactions with the Evangelical cleric had on the young student's religiosity were later recollected by the nephew in a letter to the biographers:

> It was on his recovery from that illness that I seemed to know him better than ever. It was then that my uncle's conversation seemed to make such a deep impression on his mind. He had always been a regular attendant at the services of God's house, and a regular communicant in our College Chapel. Also he had thought and read much on religious subjects. But at this time (as it appears from his own account of the matter) his religious views were greatly deepened and strengthened.[34]

It is clear from this evidence that the affect of the period of illness and recovery was fundamental in shaping Maxwell's future attitude to religion. Tayler's narrative shaped the incident in the form of an existing archetype in which Maxwell's 'conversion' consisted of a significant shift, of a kind which had been urged by the Evangelical movement from the late eighteenth-century onwards, away from his earlier broad intellectual appreciation of religion in general, and Christianity in particular, to the deeper and more committed connection with the spiritual core of his chosen faith "as grace flooded into the empty vessel of his soul".[35] This transition can also be seen as a tribute to the influence of the uncle who had spoken at some length with Maxwell during his convalescence. In practice,

[33] Campbell and Garnett, *Life*, 170.

[34] G. W. H. Tayler to Lewis Campbell, 4 March 1882, Ibid., 173.

[35] Boyd Hilton, *The Age of* Atonement*: The Influence of Evangelicalism on Social and Economic Thought 1785–1865* (Oxford University Press, 2001), 26–28.

however, and despite the elder Tayler's input, he remained, like Thomson and most other North Britons, fairly ecumenical in his theological affiliation:

> "his faith was too deep to be in bondage to any set of opinions." Scottish Calvinism was the theological system which had most historical interest for him, and most claim on his hereditary piety. [...] But that which his latest pastor has called "his deep though simple faith," was not enclosed in any system.[36]

Campbell and Garnett also noted that "in his half-speculative, half-emotional way" Maxwell had contrasted his experience in the "large and united English family" which he had found in the Tayler household with his own upbringing as the only child of a widowed father, his mother having died in 1839. It would perhaps be too facile to conjecture that, at a time when his biocultural resources were at a distinctly low ebb, the sense of belonging to a larger and more cohesive family group which he had recently experienced became associated in his *Weltmodell* with the larger fraternity of the Evangelical communion.

It should also be noted that two further significant life events provided a post-conversion Maxwell with some closure with respect to his early family life and access to alternative models of emotional attachment and community as well as a high degree of financial independence. His father died peacefully, to Maxwell's evident relief, in April 1856, leaving the family estate at Glenlair in Kircudbrightshire to James. In June 1859, he married Katherine Mary Dewar (1824–1886) and embarked on "a married life which can only be spoken of as one of unexampled devotion". In due course, he became an elder of the kirk and the "liberality" of his donations to the church "was very great".[37]

A significant consequence of Maxwell's conversion and his subsequent approach to theology, as discussed in a letter to Charles Tayler written shortly afterwards, concerned the role of free will in the acceptance of God's grace by the individual. "The ability to choose to trust God was a critical element in [F. D.] Maurice's scheme, and was also an important

[36] Campbell and Garnett, *Life*, 170–173, 32. Quotation is taken from footnote 2 on 170–171.

[37] Campbell and Garnett, *Life*, 253–257, 276, 371.

aspect of Maxwell's own understanding of his conversion experience."[38] Maxwell's conviction of the importance of every person's ability to choose "perfect freedom" rather than "bondage" was reiterated five years later in a letter written to Campbell on the occasion of the latter having taken clerical orders:

> You have to say that what men are and the nature of their actions depends on the state of their wills, and that by God's grace, through union with Christ, the contradictions and false action of those wills may be settled and solved, so that one way lies perfect freedom, and the other way bondage under the devil, the world, and the flesh, and therefore you entreat them to give heed to the things which they have heard.[39]

It is therefore unsurprising that Maxwell's key intervention in the project of the North British physicists to find a scientific basis for the less-tangible phenomena of life was an attempt to demonstrate that the exercise of free will does not require the consumption of energy. A purely biochemical basis for thought would mean that consciousness is a process which could be expected to run down and eventually fade to zero. Maxwell's preferred solution was a human body in which "Food is the mover, and perishes in the using, which the soul does not".[40] For Maxwell, this debate was not merely a matter of life and death: "the stakes were even higher: the existence of the soul and the progress of science."[41]

Maxwell in due course began to build on the idea of 'directing power' which had previously been suggested by William Thomson. At almost the same time that Thomson's interest in this topic was reawakened by a request from by his brother James, and in response to the materialist and deterministic work of Helmholtz, Alexander Bain (1810–1877), Henry Thomas Buckle (1821–1862) and William Benjamin Carpenter (1813–1885), Maxwell introduced the ideas of the shooting of a gun and

[38] Matthew Stanley, 'The Pointsman: Maxwell's Demon, Victorian Free Will, and The Boundaries of Science', *Journal of the History of Ideas*, 69, 3, July 2008: 467–491 at 471. The letter from which this evidence is derived was written by Maxwell to Charles Tayler on 8 July 1853. It can be found on pages 188–189 of Campbell and Garnett, *Life*. See also: Crosbie Smith, *The Science of Energy: A Cultural History of Energy Physics in Victorian* Britain (London: Athlone, 1998), 215–217.

[39] Maxwell to Campbell, 22 December 1857, in Campbell and Garnett, *Life*, 294.

[40] Maxwell to Campbell, 21 April 1862, in Campbell and Garnett, *Life*, 336.

[41] Stanley, 'Pointsman', 491.

the actions of a railway pointsman as analogies for the supervisory actions of the mind:

> There is action and reaction between body and soul, but it is not of a kind in which energy passes from one to the other,—as when a man pulls a trigger it is the gunpowder which projects the bullet, or when a pointsman shunts a train it is the rails that bear the thrust. But the constitution of our nature is not explained by finding out what it is not. It is well that it will go, and that we remain in possession, though we do not understand it.[42]

These two analogies provided Maxwell with starting points for thinking about ways in which free will might be exercised without requiring a significant increase in entropy. He recognised, of course, that the examples he had chosen were not entirely satisfactory: however small the directing action required to release a much greater store of energy, it could not realistically be argued to be zero. The same objection could also be raised with respect to perhaps his most well-known example of a directing power: 'Maxwell's demon'. This miniscule character was the subject of a thought experiment which he had described in an 1867 letter to Tait. Thomson, who read and annotated the letter, was responsible for giving the name 'demon' to the fictitious director of gas molecules and possessor of free will which Maxwell had envisioned.[43]

Balfour Stewart and Joseph Norman Lockyer

A further attempt to explain the exercise of free will whilst minimising or removing the requirement for the dissipation of energy took the form of a paper entitled *The place of life in a universe of energy*, which was published as the second half of a two-part article in *Macmillan's Magazine* in May 1868. The authors were Balfour Stewart (1828–1887) and Joseph Norman Lockyer (1836–1920).[44]

Stewart was a Scottish physicist and astronomer most noted for his work on radiant heat and his observation of the 'Stewart Super Flare' in

[42] Ibid., 473–474. Maxwell to Campbell, 21 April 1862, in Campbell and Garnett, *Life*, 336.

[43] Smith and Wise, *Energy*, 621–622.

[44] Balfour Stewart and J. Norman Lockyer, 'The Sun as a Type of the Material Universe: Part II The Place of Life in a Universe of Energy', *Macmillan's Magazine*, 1 May 1868: 319–327.

1861 while he was the director of the Kew Observatory. He was educated at and subsequently employed by the University of Edinburgh and was a devoted churchman who eventually became a supporter of psychical research and spiritualism.[45] His career at Kew Observatory was dogged by a series of disagreements with the chairman of the oversight committee, General Edward Sabine (1788–1883), which culminated in his resignation from his position when he feared "that the anxieties of his office might affect his health". A subsequent attempt to withdraw his resignation, following a positive interview with a would-be intercessor, was rejected by Sabine, who had already offered the job to another candidate. One of Stewart's former students, Arthur Schuster (1851–1934), felt that "The obstinacy with which Sabine pressed his own opinion is perhaps intelligible in a man who was then eighty-one years old, but there is nothing to say in extenuation of the want of generosity exhibited" in Sabine's refusal to write a testimonial in support of Stewart's application for the post of professor of Physics at Owens College in Manchester.[46]

Tait's obituary for Stewart highlighted a significant incident that befell his former colleague in 1870. Shortly after his highly unsatisfactory run-in with Sabine and just as he was taking up his new position at Owens College, Stewart was one of several casualties of "the terrible railway accident at Harrow":

> About twenty years ago Stewart met with a frightful railway accident, from the effects of which he did not fully recover. He was permanently lamed, and sustained severe injury to his constitution. From the vigorous activity of the prime of life he passed, in a few months, to grey-headed old age.[47]

The after-effects of this crash are, incidentally, reminiscent of the consequences of Thomson's curling accident and the recurring disabilities suffered by Jenkin. Stewart's recovery from his own calamity was understandably slow, preceded by five years the co-writing by Tait and Stewart

[45] 'Memoir of the late Professor Balfour Stewart, LL.D., F.R.S. By Professor A. Schuster, Ph.D., F.R.S., F.R.A.S.,' in *Memoirs and Proceedings of the Manchester Literary & Philosophical Society*, Fourth Series, vol 1 (Manchester, 1888), 253–272 at 265. Paul Charbonneau, 'Stewart, Balfour' in T. Hockey (ed.), *Biographical Encyclopedia of Astronomers* (New York: Springer Science+Business Media, 2014), 2057–2059.

[46] Arthur Schuster, *Biographical Fragments* (London: Macmillan, 1932), 209–211.

[47] 'Terrible Railway Accident at Harrow', *The Western Mail*, Monday, 28 November 1870: 2. Peter G. Tait, 'Dr. Balfour Stewart, F.R.S.', *Nature*, 29 December 1887: 202–203.

of a highly speculative but nonetheless entertaining discussion of the possible existence of eternity which will be discussed in due course.[48] Given the nature of the argument previously advanced concerning the biocultural springs of intellectual currents, it can be argued that this near-fatal accident accelerated the development of his incipient spiritual ideas, as illustrated by his initial, rather tentative, work with Lockyer.

A near contemporary parallel, which illustrates the significant effect of a similarly disturbing event, is to be found in the case of Charles Dickens (1812–1870), who "narrowly escaped death" when he was involved in a railway accident in 1865. As Jill Matus has observed, he was subsequently afflicted by a wide variety of symptoms including irrational fears, uncontrolled shaking and mental flashbacks, particularly during train journeys:

> His son Henry recalled that he got into a state of panic at the slightest jolt; Mamie [Dickens' daughter] attested that her father's nerves were never really the same again: he "would fall into a paroxysm of fear, tremble all over and clutch the arms of the railway carriage."[49]

Further indications of Dickens' frame of mind in the period after the Staplehurst crash can be found in two publications which he worked on prior to his death which coincidentally occurred five years to the day after his accident. One was a short story with supernatural overtones featuring a spectral figure whose appearances preceded a series of railway accidents (*The Signal-Man*, 1866). His final novel was pervaded with foreboding and titular dread (*The Mystery of Edwin Drood*, 1870). The plot of this unfinished book featured opium addiction, sexual obsession and murder. It opened with John Jasper's opium dream, which echoes some of those experienced earlier by Thomas De Quincey (1785–1859):

> cymbals clash, and the Sultan goes by to his palace in long procession. Ten thousand scimitars flash in the sunlight, and thrice ten thousand dancing-girls strew flowers. Then, following white elephants caparisoned in countless gorgeous colours, and infinite in number and attendants.[50]

[48] Schuster, *Biographical*, 206.

[49] Jill L. Matus, 'Trauma, Memory, and Railway Disaster: The Dickensian Connection', *Victorian Studies*, 43, 3 (Spring, 2001): 413–436 at 413–414.

[50] Charles Dickens, The Mystery of Edwin Drood (London: Chapman and Hall, 1870), 1. Only six of the planned twelve instalments were completed before Dickens' death in 1870.

Dickens couched his post-crash existential dread in the form of gothic horror fiction, a genre in which the prolific and popular author was very experienced and which constituted for him a natural mode of expression. The physicist and astronomer, Stewart, as we shall shortly see, resorted instead to imagined universes which he argued were nonetheless consistent with the tenets of existing science.

Lockyer was an amateur astronomer and civil servant from the English midlands who eventually became both an astrophysicist and an author of popular science books. He is credited with the discovery of the element helium in solar spectra but is probably best known as the editor of the journal *Nature* from 1869 to 1919. He had been, as a younger man, a Broad Church Anglican and a Christian Socialist. His first foray into scientific journalism had occurred at a relatively early age. "In 1862 his Wimbledon neighbour Thomas Hughes (1822–1896), a well-known Christian Socialist, had recruited Lockyer to help him establish a new journal called *Reader* that would allow for uninhibited discussion of British science, religion and art."[51] He was at that time seen to be close to the X-Club, whose members published a number of articles in early editions of *Nature*, but it is apparent from some of the correspondence in the columns of the journal that, at times, Lockyer had to 'hold the ring' during debates between the scientific naturalists and members of the North British group.[52]

Lockyer suffered several bouts of illness during the mid-1860s, culminating in "a more prolonged one in the middle of 1868". Following this last malady, which may have been related to what amounted to a significant setback to his civil service career at the War Office, he took a lengthy convalescence which included a walking tour in Switzerland.[53] This affliction, followed by a long break, provided the space and perhaps the motivation for co-writing with Stewart a paper in which inter alia they attempted to identify an underlying basis for free will which remained consistent with the thermodynamic principle of dissipation.

Some of the points discussed by Stewart and Lockyer in *The place of life* echoed earlier arguments made by Thomson and Maxwell. Although

[51] Melinda Baldwin, 'The shifting ground of *Nature*: establishing an organ of scientific communication in Britain, 1800–1900', *Hist. Sci.* 1 (2012): 125–152 at 128. *Reader* was published from 1862 until 1867. Arthur J. Meadows, *Science and controversy: a biography of Sir Norman Lockyer* (London: Macmillan, 1977), 9.

[52] Barton, 'Scientific Authority'.

[53] Meadows, *Science*, 10–12.

energy cannot be created, it can be transformed, and the nature and timing of such changes can, in principle, be managed by an infinitesimally tiny quantities of consciously directed energy. The particular example used in the Stewart and Lockyer article was the employment of "less than any assignable amount" of energy in the mind of an actor who pulls the trigger of a gun, the end result of which might be that "if well pointed, it may explode a magazine,—nay, even win an empire."[54] They proposed "two hypotheses of the action of life" which they initially appeared to decline to choose between:

> The first of these is the materialistic hypothesis, which denies the existence of life as a principle apart from matter; while the other allows the existence of an independent principle, but assumes its action to take place through the medium of a machine of infinite delicacy, so that by a primordial impulse of less than any assignable amount a finite and visible outcome is produced. These are two alternatives, and it is not within our province to attempt to decide between them. The battle must be fought in other pages than ours, and by other weapons than those which we can produce.[55]

The rather tautological argument in which the discharge of a small, but not quite zero, quantum of energy acts as a sort of proxy for free will and consciousness is clearly, despite the disclaimer, their preferred solution. A little farther on, they dropped a heavy "hint to those who are inclined to adopt that theory of life which demands an infinite delicacy of construction" by extending their hypothesis of (almost) energy-free interactions to an even more tenuous concept of "action at a distance". By this term, they meant distance in both time and space, thus providing a possible basis for spiritual phenomena which they were disinclined to endorse strongly but which does provide a signpost towards Stewart's later collaboration with Tait and his engagement with spiritualism.[56]

Further Contributions from Jenkin and Maxwell

Stewart and Lockyer's approach to possible explanations of the underlaying basis of free will was rather more cautious than that of Fleeming Jenkin who wrote a review, also published in 1868, of the English translation of

[54] Stewart and Lockyer, 'The Sun', 324.
[55] Ibid., 326.
[56] Stewart and Lockyer, 'The Sun', 327.

'*De Rerum Natura*' by Hugh Andrew Johnstone Munro (1819–1885) which was well received by its author: "Munro himself owned that he had found instruction in the paper on Lucretius". This article provided further support for Thomson and Maxwells' concepts of the non-material operation of free will. Specifically, Jenkin "offered a plausibility argument" to the effect that if the will "involving a power in man beyond that of atoms" acted continually at right angles to an object that it wished to influence, then no energy would be lost in the process:

> The will, if it so acted, would add nothing sensible to nor take anything sensible from the energy of the universe. The modern believer in free-will will probably adopt this view, which is certainly consistent with observation, although not proved by it.[57]

Jenkin here is equating the operation of free will with the idea that atoms can randomly swerve without being acted on by any external force. "The motive for devising the curious doctrine that atoms might swerve now and then from the straight path without being acted upon by other atoms, was as Mr. Munro observes, undoubtedly the desire to devise an explanation of Free-will."[58]

Maxwell, in an essay written in 1873, endorsed the association of the operation of free will with Lucretius' swerve by positing a hypothetically infinitesimal tipping point as an alternative to his earlier 'pointsman':

> Our free will at the best is like that of Lucretius's atoms, which at quite uncertain times and places deviate in an uncertain manner from their course. In the course of this our mortal life we more or less frequently find ourselves on a physical or moral watershed, where an imperceptible deviation is sufficient to determine into which of two valleys we shall descend. The doctrine of free will asserts that in some such cases the Ego alone is the determining cause.[59]

This ongoing dilemma of the operation of the will without the consumption of a finite amount of energy was almost, but not quite, resolved

[57] Smith and Wise, *Energy*, 628. Fleeming Jenkin, 'Lucretius and the atomic theory', in Colvin and Ewing, *Fleeming Jenkin*, vol 1, 177–214 at 192–193.

[58] Ibid., 192.

[59] 'Science and Free Will', 11 February 1873, in Campbell and Garnett, *Life*, 434–444 at 441.

at a later date by the work of a French hydrodynamicist called Joseph Boussinesq (1842–1929), who was endeavouring to demonstrate the indeterminacy of fluid flows in relation to the vortex atoms on which Thomson and Stewart had also been working:

> In most of the former methods, Dr. Balfour Stewart's &c. there was a certain small but finite amount of travail decrochant or trigger-work for the will to do. Boussinesq has managed to reduce this to mathematical zero, but at the expense of having to restrict certain of the arbitrary constants of the motion to mathematically definite values, and this I think will be found in the long run very expensive. But I think Boussinesq's method is a very powerful one against metaphysical arguments about cause and effect and much better than the insinuation that there is something loose about the laws of nature, not of sensible magnitude but enough to bring her round in time.[60]

As it happened, Maxwell's letter, in which he saw the glimmerings of a solution to his quandaries about the existence of non-dissipating psychic components of the soul, including the elusive but essential property of free will, was written just before he had "begun to be troubled by difficulties in swallowing—the first significant symptom of the disease which was to prove fatal". He died in November 1879, after several months of this serious illness which was accompanied by severe pain.

He was visited daily by his parish priest who subsequently described the continuing firmness of his faith and composure in the face of death which were stimulated and informed by his reading of the religious writers who had provided spiritual sustenance for his continued close allegiance to Christianity: "Milton, Keble, Newman, Wesley, George Herbert the latter his chief favourite". The cleric reported that Maxwell continued to subscribe wholeheartedly to the primary tenets of his chosen theology, spurred on by awareness of his forthcoming demise:

> his illness drew out the whole heart and soul and spirit of the man; his firm and undoubting faith in the Incarnation and all its results; in the full sufficiency of the Atonement; in the work of the Holy Spirit. He had gauged and fathomed all the schemes and systems of philosophy, and had found them

[60] Maxwell to Francis Galton, 26 February 1879, Francis Galton Papers, The Library, University College, London. There is a copy of this letter in Cambridge University Library, MS Add.7655/III/a/27.

utterly empty and unsatisfying—"unworkable" was his own word about them—and he turned with simple faith to the Gospel of the Saviour.[61]

The Unseen Universe

Maxwell's letter to Francis Galton (1822–1911) describing the work of Boussinesq had been written in part as a riposte to Tyndall's *Belfast Address*. Another reaction emanating from the North British group which was clearly triggered by the *Address* was the attempt by Balfour Stewart and Peter Guthrie Tait to pose a spiritual alternative to what they saw as Tyndall's materialist manifesto. That this was an objective shared with a number of North British physicists and that it was to some extent a coordinated response which had been taking shape for some time can be understood from the final line in a letter from Stewart to the engineer David Forbes (1849–1936, son of James David Forbes) written in March 1874: "I shall wait expectantly for your answer to Tyndall—we are also preparing a file for him here about which you will hear in time."[62]

Tait and Stewart jointly wrote a book called *Unseen Universe or Physical Speculations on a Future State*, which first saw the light of day in 1875. Schuster was firmly of the opinion that most of the content in the first edition had originated from Stewart. Tait's initial contribution was limited to a preface which "was almost entirely taken up with an attack on John Tyndall" and had to be completely rewritten with the exception, perhaps, of the first paragraph which Schuster surmised was probably part of Tait's original draft:

> Forgetful of the splendid example shown by intellectual giants like Newton and Faraday, and aghast at the materialistic statements nowadays freely made (often professedly in the name of science), the orthodox in religion are in somewhat evil case.[63]

Stewart and Tait's objective was to combine the temporally limited visible universe, which had been sentenced to death by thermodynamics, with an infinite but invisible creation designed and built by a "Supreme

[61] Campbell and Garnett, *Life*, 411–416.

[62] Balfour Stewart to David Forbes, 10 March 1874, Courtesy of the University of St Andrews Libraries and Museums, msdep7, 1874, letter 7.

[63] Schuster, *Biographical*, 214. Balfour Stewart and Peter G. Tait, *The Unseen Universe: or Physical Speculations on a Future State* (London: Macmillan, 1875), vii.

Governor" or "Absolute One" who happened to also be the God of Christian revelation. They thereby endeavoured to reconcile the new ideas promulgated by science with pre-existing beliefs derived from religious texts:

> The truth is, that science and religion neither are nor can be two fields of knowledge with no possible communication between them. Such a hypothesis is simply absurd. There is undoubtedly an avenue leading from the one to the other, but this avenue is through the unseen universe.[64]

Christianity, in their view, was unique in that it possessed texts and revelations which demonstrated a greater understanding of the true nature of the universe than do the other religions which they had briefly and somewhat partially surveyed in their early chapters:

> If then we regard the universe from this point of view we are led to a scientific conception of it which is, we have seen, strikingly analogous to the system which is presented to us in the Christian religion. For not only are the nebulous beginning and fiery termination of the present visible universe indicated in the Christian records, but a constitution and power are therein assigned to the Unseen Universe strikingly analogous to those at which we arrive by a legitimate scientific process.[65]

Their rationale for interconnecting an infinite, external and invisible universe with the more mundane construct which is all that limited human senses can perceive was derived somewhat adventitiously from the first law of thermodynamics, which they rebranded as the "Principle of Continuity" and which they claimed meant that matter and energy cannot be created *ex nihilo* but must always have existed and cannot be destroyed. The continuing existence of everything which had ever existed was in turn the mechanism by means of which they managed to justify the parallel concept of immortality for sentient beings. In their expanded cosmos, the seen and unseen universes are connected via the medium of the ether which is responsible for the transmission of thoughts from the former to latter, thus creating and maintaining a set of 'shadow' souls in this invisible place.

This perfectly preserved parallel universe which provides a safe harbour for the soul and the means of indefinite preservation of the status quo can

[64] Ibid., 60, 168, 211.

[65] Stewart and Tait, *Unseen*, 1875, 210.

also be read as an otherworldly embodiment of the right-wing political views of Tait whose "general outlook upon human affairs was fundamentally conservative". Despite his vehement opposition to Tyndall's materialism he would have in other respects concurred with his opponent on some more mundane topics, including having "had a deep distrust of Mr Gladstone as statesman and legislator".[66]

In a later version of *Unseen Universe*, Stewart and Tait expanded the scope of their thesis significantly by alluding to the work of Thomas Young (1773–1829), who had proposed the existence of a series of further invisible universes. Young's 'Russian doll' concept was built on a supposition that our three-dimensional universe sits within a higher-order four-dimensional universe. The invisible fourth dimension was then supposed to exist within a five-dimensional cosmos and so on, presumably *ad infinitum*:

> It seems therefore natural to believe that the analogy may be continued still further, until it rises into existences absolutely immaterial and spiritual. We know not but that thousands of spiritual worlds may exist unseen for ever by human eyes; nor have we any reason to suppose that even the presence of matter, in a given spot, necessarily excludes these existences from it.[67]

As a logical consequence of the supposed additional dimensions and also in response to a somewhat-negative review by William Kingdon Clifford (1845–1879), they then adduced the idea that higher-order universes could intrude themselves into lower-order entities "so the matter of our present universe may be regarded as produced by mere rents or cracks in that of the Unseen".[68]

These openings in the surface of our world were supposed to give rise to portals into the invisible sphere which the authors also equated with heaven in much the same way as lakes and ponds were regarded in Celtic mythology as gateways into the underworld.[69] The authors sought to base this speculation on a biblical source by alluding to the apostle Paul and his

[66] Knott, *Tait*, 37.

[67] Balfour Stewart and Peter G. Tait, *The Unseen Universe: or Physical Speculations on a Future State* (London: Macmillan, 1886), 160–161.

[68] William K. Clifford, 'The Unseen Universe', *Fortnightly Review*, May 1865–June 1934; Jun 1875; 17, 102: 776–793. Stewart and Tait, *Unseen*, 1886, 221.

[69] John T. Koch, *Celtic Culture: A Historical Encyclopedia*, (Santa Barbara CA: ABC-Clio, 2006), 1671.

vision of a "third heaven".[70] They proceeded to employ such 'magic doors' to account for prophecies and miraculous events in the Bible, thus countering earlier challenges to the credibility of such narratives: "Therefore miracles were not discontinuities or examples of special providence but subject to law."[71]

The idea of connecting doors between dimensions would have particularly appealed to Stewart, whose interest in spiritualism was to lead to him becoming a founder of the Society for Psychical Research (SPR), eventually taking on the role of president. Despite a colleague's observation that Stewart's attitude was one of "impartial and philosophical inquiry, and that no subject is beneath the attention of scientific men", it is not difficult to trace a connection between his close brush with death and his subsequent engagements with both *The Unseen Universe* and "the many new deaths discovered in the practices of spiritualism and psychical research".[72]

Stewart and Tait had, at least to their own satisfaction, successfully defended some key tenets of the Christian faith, including most crucially an eternal heaven in which the immortal souls of believers could survive, from two contemporary threats. The first of these was the attempt by agnostic scientific naturalists to remove all requirements for divine management from the developmental laws and processes which shape the world as we now see it. The second was the threat to infinite time which had arisen from their fellow physicists, some of whom professed the Christian faith. They had pronounced a doom on the visible universe by the introduction of the inevitability of entropic decay. Or as Darwin put it in his autobiography:

> With respect to immortality, nothing shows me how strong and almost instinctive a belief it is, as the consideration of the view now held by most physicists, namely that the sun with all the planets will in time grow too cold for life, unless indeed some great body dashes into the sun and thus gives it fresh life.—Believing as I do that man in the distant future will be a far more perfect creature than he now is, it is an intolerable thought that he and all

[70] Stewart and Tait, *Unseen*, 1886, 221. The original Pauline reference is 2 Corinthians 14:2–4.

[71] Hilton, Atonement, 367.

[72] Schuster, 'Memoir', 265. Rhodri Hayward, *Resisting History: Religious Transcendence and the Invention of the Unconscious* (Manchester University Press, 2008), 21.

other sentient beings are doomed to complete annihilation after such long-continued slow progress.[73]

The book turned out to be rather more popular than perhaps even its authors had anticipated, eventually running to four editions and nine reprints. It was still in print in 1890.[74] One consequence of the success of *Unseen Universe* was its adoption as a kind of spiritual 'call to action' by the psychologist William James (1842–1910), who had reviewed the book shortly after its initial appearance. Prompted by the optimistic tone of Stewart and Tait's work and its apparently elegant resolution of the potentially conflicting findings of religion and science, James concluded that there was "a 'duty' to believe, holding that belief in a transcendent realm was not only permissible but something which one may be duty bound to hold if it would, for the believer, be a source of commendable action or peace of mind".[75] He went on to turn this uncritical acceptance of some of the core tenets of the book into the fundamental thesis of a later lecture and book entitled *The Will to Believe* in which he argued that religious belief in the absence of evidential support is permissible in certain circumstances.[76] This interpretation would have been somewhat at odds with the views of the original authors in that they seem to have convinced themselves that sufficient evidence existed for their proposed regression of universes as the underpinning of faith.

The North British Group and the Immaterial Universe

The main argument made in preceding chapters has been the contingency of changes in theological, philosophical and metaphysical perspectives which can be traced, at least in part, to affective life events which precipitated a significant reappraisal of personal epistemologies. It is, on the basis

[73] Nora Barlow, *The Autobiography of Charles Darwin 1809–1882. With the original omissions restored. Edited and with appendix and notes by his grand-daughter, Nora Barlow* (London: Collins, 1958).

[74] Knott, *Tait*, 365.

[75] William James, 'Review of *The Unseen Universe*', *Nation*, 20 (1875): 366–367. Hunter Brown, *William James on Radical Empiricism and Religion* (University of Toronto Press, 2000), 29.

[76] William James, *The Will to Believe and other Essays in Popular Philosophy* (New York: Longmans Green, 1897).

of this discourse, unsurprising that such connections have again been found in the narrative arcs of the scientists whose stories we have surveyed in this chapter. These *Weltmodell* disruptions have been culled from the biographies of a selection of subjects who shared a broadly anti-materialist position. At first sight, there may appear to be more than a suspicion in the cases of both northern physicists and southern biologists that their epistemological views can be traced to their early exposure to either fundamental Calvinism, in the form of Scottish Presbyterianism, on the one hand or the more reflexive strains of Anglicanism and Unitarianism on the other. As is often the case, however, if we look past the superficial evidence, the picture is rather more nuanced.

There appear to have been several factors which inspired the North British physicists to produce their less materialistic, and occasionally overtly religious, ruminations on the nature of life, the universe and everything. Although they for the most part shared a broadly similar religious perspective, this was not the only factor which drove their *Weltmodelle*. They had all been wholly or partly raised in Scotland and presumably felt a degree of competition with metropolitan English scientists who held broadly agnostic views. The threat of intellectual dominance in their chosen profession must also have been a cause for rivalry and a spur to competition for the Caledonian contingent. Traumatic disruptions and long-term illnesses clearly played some part in the timings of their responses. The affective and occasionally existential crises which were prompted by these events were, at least to some extent, the stimuli which crystallised their responses. Such setbacks also acted as breaks from the day-to-day considerations of more mundane matters and an opportunity for contemplation of more abstract, and yet also quite fundamental, thoughts. Thomson's broken leg, for example, "prevented his experimenting for a whole winter".[77]

Balfour Stewart, whose religious affiliation remained broad, culminating in a general interest in spiritualism, had already begun to think about the nature of free will and consciousness when he co-wrote *The place of Life* with a convalescent Lockyer in 1868. Two years later, he had his leg broken in a serious railway accident which had also caused a number of fatalities amongst his fellow passengers.[78] It preceded the writing by Stewart and his colleague Tait of their highly speculative but nonetheless

[77] Peter Tait to Arthur Cayley, 4 February 1886, quoted in Knott, *Tait*, 189.
[78] 'Railway Accident', 2.

entertaining tome on science, religion and immortality by approximately five years. His interest in, and pursuit of, these topics may well have been further spurred by his injury and its after-effects, although it is also clear that the proximate cause of the writing of *Unseen Universe* was the sense of provocation caused by Tyndall's *Address*.

Fleeming Jenkin's scientific career encompassed a variety of fields. He acted as a populariser of Thomson's work on terrestrial longevity, producing more understandable, but also more polemical, versions of the arguments which his fellow physicist had made. Jenkin was perhaps the least religiously inclined member of the North British group, and his involvement in the chronology and free will debates seems to have sprung from his membership of, and loyalty to, the in-group formed by a particular professional community as much as from his theological conviction. An interesting aspect of Jenkin's interventions, however, is that they did not follow immediately after either the publication of Darwin's thesis in 1860 or the counterarguments which emanated from Thomson in 1862, but actually began in 1867, two years after a night of considerable anxiety and, perhaps more importantly, the advent of a frustrating disability which echoes those of Thomson and Stewart.

James Clerk Maxwell, having undergone a classical conversion experience, had thereby gained sources of religious inspiration for his speculative ideas about the seat of free will and consciousness. Being less dependent on his scientific work as a means of financial support than some of his west coast peers, he found that he tended to "get fonder of metaphysics" as his career progressed. As a consequence, he favoured academic purity over practical application:

> I find I get fonder of metaphysics and less of calculation continually [...] Experiment furnishes us with the values of our arbitrary constants, but only suggests the form of the functions. Afterwards, when the form is not only recognised but understood scientifically, we find that it rests on precisely the same foundation as Euclid does, that is, it is simply the contradiction of an absurdity, out of which may we all get our legs at last![79]

He was subsequently acknowledged by Albert Einstein (1879–1955) and others as having produced some of the most important theoretical conclusions of the nineteenth century. Although there was a hiatus

[79] James Maxwell to R. B. Litchfield, 4 July 1856, in Campbell and Garnett, *Life*, 261.

between his death in 1879 and the wider recognition of the validity of his work on the unity of different forms of electromagnetic radiation, the equations with which he described their behaviour remained "relativistically" valid during and after the rewriting of the existing laws of physics by Einstein, who was of the opinion that

> we may say that, before Maxwell, Physical Reality, in so far as it was to represent the processes of nature, was thought of as consisting in material particles, whose variations consist only in movements governed by partial differential equations. Since Maxwell's time, Physical Reality has been thought of as represented by continuous fields, governed by partial differential equations, and not capable of any mechanical interpretation.[80]

Such a conclusion was of course fundamentally attuned to the desires of the North British group to augment the materialist construction of the seen universe with the immaterial and spiritual components which remained invisible and eternal but which they sensed were in danger of being overlooked by those of their metropolitan counterparts.

The case studies presented in this chapter, and in this book as a whole, have demonstrated the existence of potential interconnections between apparently objective cognitive processes with the anticipated impact of perceived challenges to the integrity of the self. To reiterate the major argument outlined above, significant changes in the development of the physical sciences and overlapping epistemologies were sometimes contingent on preceding instances of exceptional affect, accompanied at times by the unpredictable and largely undocumented effects of psychotropic substances, which drove contingent changes in the subjects' *Weltmodell*. The extent of the resulting change depended on the severity of the stimulus and the closeness of the object of an event to the individual in question: threats to the continuity of the subjects themselves being evidently the most influential. We cannot presume to interpolate ourselves into nineteenth-century minds or bodies, nor can we necessarily read forward from such relatively distant *Weltmodelle* to contemporary mind-sets, but it does seem clear that we should add to the armoury of the historian of science a further component of the formation of ideas: the connection

[80] Paul J. Nahin, 'Maxwell's grand unification', *IEEE Spectrum*, March 1992: 45. Albert Einstein, 'Maxwell's influence on the development of the concept of physical reality' in *James Clerk Maxwell: A Commemoration Volume 1831–1931* (Cambridge: The University Press, 1931), 66–73 at 71.

between existentially freighted life events and significant alterations in the subject's selection of espoused theories, performed selves and observed behaviours. Applying this mode of interpretation to the broader field of nineteenth-century scientific developments in general, we shall achieve a greater understanding of some significant and previously overlooked undercurrents which have influenced the intellectual history of that era. They are still relevant today and will continue to be so into the future.

Bibliography

Archive Sources

Correspondence between Edward Clodd and Richard Anthony Proctor can be found in the Clodd Correspondence, Brotherton Library Special Collections, Leeds University, Leeds LS2 9JT.

Most of the correspondence of John William Frederick Herschel quoted here was accessed in the library of The Royal Society, 6–9 Carlton House Terrace, London SW1Y 5AG.

Correspondence between John William Frederick Herschel and John William Whittaker can be found in the Library of St John's College, Cambridge CB2 1TP.

The correspondence and pamphlets of the 3rd Earl of Rosse (William Parsons) is held in the archives of Birr Castle, Birr, Co. Offaly, Republic of Ireland.

Family correspondence of William Thomson (1st Baron Kelvin) can be found in the archives of Cambridge University Library, West Rd, Cambridge CB3 9DR.

Miscellaneous correspondence of Thomas Chalmers, William Thomson and Balfour Stewart is archived by the University of St Andrews Library, North Street, St Andrews KY16 9TR.

The digitised correspondence of Charles Darwin can be found online at: www.darwinproject.ac.uk.

H. Carlton, *Cosmology and the Scientific Self in the Nineteenth Century*, https://doi.org/10.1007/978-3-031-05280-4

Bibliography of Primary Sources

Biblical references and quotations are taken from the King James Version, printed by Eyre and Spottiswoode for the British and Foreign Bible Society (London, 1857).

'Appendix on the Age of the Sun', added 15 August 1854, *Transactions of the Royal Society of Edinburgh*, vol XXI (Edinburgh: Robert Grant, 1857), 63–80.

'Argument to prove the truth of the Christian revelation', *The Athenaeum*, 26 April 1834; 339: 311–312.

'Astronomical Notes', *The Athenaeum*; 1 March 1890; 3253: 279.

'Astronomy', *The Scotsman*, 28 March 1835, 1.

'Dr. Nichol on the system of the world (Book Review)', *The Spectator*, 12 September 1846; 19, 950: 882–883.

'Dr. Nichol's Architecture of the heavens (Book Review)', *The Spectator*, 22 July 1837; 10, 473: 689–690.

'Journal of Science, Inventions, and improvements', *The Critic*, 1844–1847; Apr 11, 1846; 3, 67: 405.

'Mr J. P. Nichol.', *The Scotsman*, 24 October 1835, 3.

'Our Weekly Gossip', *The Athenaeum* dated 17 October 1846, 1072.

'Printed address by J. Carlile to the Roman Catholic members of the Mechanics Institute of Parsonstown', 25 October 1845, Birr Castle MSS J 9.

'Prof. Proctor: Changed views as to Science and Religion', *Chicago Daily Tribune*, 10 November 1875, 7.

'Prof. Proctor: Germs of his present views in his earlier writings', *Chicago Daily Tribune*, 14 November 1875, 3.

'Professor Nichol', *The Gentleman's Magazine: and historical review*, Nov 1859; 207: 536.

'Professor Nichol', *The Scottish Review*; January 1860: 1–17.

'Professor Nichol's lectures on astronomy', *The Manchester Guardian*, 15 May 1847, 4.

'Professor Nichol's lectures on astronomy', *The Manchester Guardian*, 22 May 1847, 10.

'Professor Nichol's lectures on recent astronomical discoveries', *The Manchester Guardian*, 12 May 1847, 6.

'Professor Nichol's lectures on recent astronomical discoveries', *The Manchester Guardian*, 26 May 1847, 3.

'Professor Nichol's new works', *Tait's Edinburgh magazine*, Mar 1848; 15, 171: 200.

'Professor Nichol's Views of the Architecture of the Heavens.', *Chambers' Edinburgh journal*, 29 July 1837: 210–211.

'Richard A. Proctor Dead', *The New York Times*, 13 September 1888, 1.

'Suicide of Eyton Tooke, Esq.', *The Observer*, 31 January 1830, 3.

'Terrible Railway Accident at Harrow', *The Western Mail*, Monday, 28 November 1870: 2.
'The Origin of Man', *The New York Times*, 5 September 1874, 1–4.
'The Progress of Science', *The New York Times*, 7 September 1874, 3.
'Travels in Search of Lord Rosse's Telescope', *Chambers' Edinburgh Journal*, 26 December 1846; 156: 401–404.
2nd Earl of Rosse, *An Argument to prove the Truth of the Christian Revelation* (London: John Murray, 1834).
3rd Earl of Rosse, 'Observations on the nebulae' in Parsons (ed.), *scientific*, 109–124.
Airy, George B., 'Address', *British Association Report* (1851), xxxix–liii.
Allen, Grant, 'Our Ancestors: I – The Stone Age Men', *Knowledge*, vol 1, 24 February 1882: 351–352.
Andrew D., *A History of the Warfare of Science with Theology in Christendom*. 2 volumes (London: Macmillan, 1897).
Ball, Robert S., *Great Astronomers* (London: Isbister, 1895).
Barlow, Nora, *The autobiography of Charles Darwin 1809–1882. With the original omissions restored. Edited and with appendix and notes by his grand-daughter, Nora Barlow* (London: Collins, 1958).
Barry, William F., 'Mr. Tyndall and Contemporary Thought.' *The Dublin Review*, Oct 1876, 27, 54: 431–469.
Brown, Samuel, 'The History of Science (1846)', *Lectures on the Atomic Theory and Essays Scientific and Literary*. vol 1 (Edinburgh: Thomas Constable, 1858), 299–344.
Campbell, Lewis and William Garnett, *The Life of James Clerk Maxwell* (London: Macmillan, 1882).
Chalmers, Thomas, *A Series of Discourses on the Christian Revelation viewed in connection with the Modern Astronomy* (Glasgow: John Smith and Son, 1817).
Chalmers, Thomas, *Institutes of Theology*, vol II (Edinburgh: Sutherland and Knox, 1849).
Chalmers, Thomas, *On Natural Theology*, Volume 1 (New York: Robert Carter & Brothers, 1857).
Chalmers, Thomas, *On Political Economy in connection with the Moral State and Moral Prospects of Society* (Glasgow: William Collins, 1832).
Chalmers, Thomas, *On the Power, Wisdom and Goodness of God, as manifested in the adaptation of external nature to the Moral and Intellectual Constitution of Man* (1833 repr. Glasgow: William Collins, 1839).
Chambers, Robert, *Vestiges of the Natural History of Creation* (London: John Churchill, 1844).
Clark, John W., *Old Friends at Cambridge and Elsewhere* (London: Macmillan, 1900).

Clerke, Agnes M., 'Herschel, John William Frederick', *Dictionary of National Biography* (London: Smith, Elder & Co, 1885–1900), vol xxvi, 263–268.

Clerke, Agnes M., *A Popular History of Astronomy during the Nineteenth Century*, 2nd Edition (Edinburgh: A. & C. Black, 1887).

Clifford, William K., 'The Unseen Universe', *Fortnightly Review*, May 1865–June 1934; Jun 1875; 17, 102: 776–793.

Clodd, Edward, 'The Birth and Growth of Myth', *Knowledge*, 4, 21 September 1883: 182–184.

Clodd, Edward, *Memories* (London: Chapman and Hall, 1916).

Clodd, Edward, *The Childhood of the World: a simple account of man in early times* (London: Macmillan, 1873).

Cobbe, Frances P., *Life of Frances Power Cobbe by herself*, vol 1 (Boston: Houghton, Mifflin, 1895).

Coleridge, Samuel T., 'Table Talk' (22 February 1834) in W. G. T. Shedd (ed.), *The Complete Works of Samuel Taylor Coleridge*, vol VI (New York: Harper and Brothers, 1884), 502–503.

Colvin, Sidney and J. Alfred Ewing (eds.), *Papers, Literary, Scientific etc. by the late Fleeming Jenkin, F.R.S., LL.D.* 2 volumes (London: Longmans, Green, 1887).

Conway, Moncure D., *The Life of Thomas Paine*, vol II (New York: G. P. Putnam's Sons, 1892), 128–132.

Coutts, James, *A History of the University of Glasgow; from its foundation in 1451 to 1909* (Glasgow: James MacLehose and Sons, 1909).

Darwin, Charles, *On the Origin of Species by means of Natural Selection, or the Preservation of Favoured Races in the struggle for life* (London: John Murray, 1859).

Darwin, Charles, *On the Origin of Species by means of Natural Selection, or the Preservation of Favoured Races in the struggle for life* (New York: D. Appleton, 1860).

Darwin, Francis (ed.), *The life and letters of Charles Darwin, including an autobiographical chapter*, vol 1 (London: John Murray, 1887).

Darwin, Francis and Albert C. Seward (eds.), *More Letters of Charles Darwin*, vol II (London: John Murray, 1903).

de Maupassant, Guy, 'Dreams' originally published as 'Rêves' in *Le Père Milon* (Paris: Paul Ollendorff, 1899).

De Quincey, Thomas, 'System of the heavens as revealed by Lord Rosse's telescopes', *Tait's Edinburgh magazine*, Sep 1846; 13, 153: 566–579.

De Quincey, Thomas, *Confessions of an English opium-eater*, Second Edition (London: Taylor and Hessey, 1823).

De Quincey, Thomas, *Confessions of an English opium-eater*, Volume one in a series of De Quincey's works (Edinburgh: Adam and Charles Black, 1862).

Dick, Thomas, *The Philosophy of a Future State* (New York: G. & C. & H. Carvill, 1829).

Dick, Thomas, *The Practical Astronomer* (London: Seeley, Burnside and Seeley, 1845).
Dick, Thomas, *The Sidereal Heaven and other subjects connected with Astronomy* (New York: Harper & Brothers, 1840).
Dickens, Charles, The Mystery of Edwin Drood (London: Chapman and Hall, 1870).
Draper, John W., *History of the Conflict between Religion and Science* (New York: Appleton, 1875).
Duncan, David, *The Life and Letters of Herbert Spencer* (London: Methuen, 1908).
Dunn, Samuel, 'Reminiscences of Dr. Chalmers', *The Wesleyan-Methodist magazine*, Nov. 1847, 3: 1077–1099.
Eberty, Felix, *The Stars and the Earth* (London: H. Bailliere, 1846).
Everett, Joseph D., 'Letter', *Nature*, 20 April 1876, 486.
Gilfillan, George, *A Gallery of Literary Portraits* (Edinburgh: William Tait, 1845).
Gilfillan, George, *A Second Gallery of Literary Portraits* (Edinburgh: James Hogg, 1850).
Gordon, H. Laing, *Sir James Young Simpson and Chloroform (1811–1870)* (London: T. Fisher Unwin, 1897).
Hall, Spencer T., *Biographical Sketches of Remarkable People, chiefly from Personal Recollection, Miscellaneous Papers and Poems* (London: Simpkin, Marshall, 1873).
Hanna, William, *Memoirs of the Life and Writings of Thomas Chalmers D.D., LL. D.*, vol I (Edinburgh: Sutherland and Knox, 1849).
Hanna, William, *Memoirs of the Life and Writings of Thomas Chalmers D.D., LL. D.*, vol II (Edinburgh: Sutherland and Knox, 1850).
Hanna, William, *Memoirs of the Life and Writings of Thomas Chalmers D.D., LL. D.*, vol IV (Edinburgh: Thomas Constable, 1852).
Hardy, Thomas, *Two on a Tower* (London: Macmillan, 1923).
Herschel, John F. W., 'Address to the British Association for the Advancement of Science, June 19. 1845' in *Essays from the Edinburgh and Quarterly Reviews, with addresses and other pieces* (London: Longman, Brown, Green, Longmans & Roberts, 1857), 634–682.
Herschel, John F. W., 'Sir John Herschel's reply to Dr. Robinson', *The Athenaeum*; Nov 4, 1843; 836: 983–984.
Herschel, John F. W., *A Preliminary Discourse on the Study of Natural Philosophy* (London: Longman, Rees, Orme, Brown, Green & Longman, 1832).
Herschel, John F. W., *A Treatise on Astronomy* (London: Longman, Reese, Orme, Brown, Green & Longman, 1833).
Herschel, John F. W., *Outlines of Astronomy* (Philadelphia: Lea & Blanchard, 1849), 506–513.
Hodder, Karen (ed.), *The Works of Alfred Lord Tennyson* (Ware: Wordsworth Editions, 1994).
Homer, (tr. Eric V. Rieu), The Odyssey (London: Guild Publishing, 1987).

Huggins, Margaret L., *Agnes Mary Clerke and Ellen Mary Clerke: An Appreciation* (Printed for Private Circulation, 1907).

Hume, David, *An Enquiry Concerning the Human Understanding* (1777 repr. Oxford: Clarendon, 1894).

Hume, David, *Dialogues Concerning Natural Religion* (1779 repr. in Dorothy Coleman (ed.), *Dialogues Concerning Natural Religion: And Other Writings* (Cambridge University Press, 2007), 1–102.

Hutton, James, *Theory of the earth, with proofs and illustrations. In four parts.* vol 1 (Edinburgh: William Creech, 1795).

James, William, 'Review of *The Unseen Universe*', *Nation*, 20 (1875): 366–367.

James, William, *The Will to Believe and other Essays in Popular Philosophy* (New York: Longmans Green, 1897).

James, William, The Varieties of Religious Experience: a study in human nature, being the Gifford lectures on natural religion delivered at Edinburgh in 1901–1902 (New York: Longmans, Green, 1902).

Jenkin, Fleeming, 'Lucretius and the atomic theory' in Colvin and Ewing (eds.), *Papers*, vol 1, 177–214.

Johnson, William (ed.), *Science and Revelation; A Series of lectures in reply to the theories of Tyndall, Huxley, Darwin, Spencer, etc.* (Belfast: William Mullan, 1875).

King, Agnes G., *Kelvin the man* (London: Hodder and Stoughton, 1925).

King, Elizabeth T., *Lord Kelvin's early home* (London: Macmillan, 1909).

Kleineke, Hannes, "'Morton's Fork'? – Henry VII's 'Forced Loan' of 1496'", *The Ricardian*, xiii (2003): 315–327.

Knight, William A., *Memoir of John Nichol, Professor of English Literature in the University of Glasgow* (Glasgow: James Maclehose, 1896).

Knott, Cargill G., *Life and Scientific Work of Peter Guthrie Tait* (Cambridge University Press, 1911).

Laplace, Pierre S., (tr. J. Pond), *The System of the World* (London: Richard Phillips, 1809) Volume II, Book V.

Lightman, Bernard V., *Victorian popularizers of science: designing nature for new audiences* (University of Chicago Press, 2007).

Macgregor, Duncan C., 'Some Unpublished Letters of Dr. Chalmers', in Henry v. d. B. Copeland (ed.), *The Wimbledon & Merton Annual* (Wimbledon: Edwin Trim, 1903): 107–118.

MacLehose, James, *Memoirs and Portraits of One Hundred Glasgow Men* (Glasgow: James MacLehose & Sons, 1886).

Martineau, James, 'Modern Materialism: Its Attitude towards Theology', *The Contemporary Review*; 1 December 1875, 27: 323–346.

Maxwell, James C., 'Molecules', *Nature*, 8, 25 September 1873: 437–441.

Meadley, George W., *Memoirs of William Paley, D.D.* (Sunderland: J. Graham, 1809).

Mineka, Frances E. (ed.), *The Earlier Letters of John Stuart Mill 1812–1848 (Volume XIII of the Collected Works of John Stuart Mill)* (University of Toronto Press, 1963).

Mitchel, Frederik A., *Ormsby MacKnight Mitchel: Astronomer and General. A Biographical Narrative* (Boston: Houghton, Mifflin, 1887).

Mitchel, Ormsby M., *Popular Astronomy* (London: George Routledge and Sons, 1860).

Nichol, John P., 'ART. I.-On Political Economy, in Connexion with the Moral State and Moral Prospects of Society', *Westminster Review*, Jul 1832; 17, 33: 1–33.

Nichol, John P., 'Art. VI. State of Discovery and Speculation Concerning the Nebulae', *London and Westminster Review*, Jul 1836; 3, 2: 390–409.

Nichol, John P., 'Dr. Chalmers', *Tait's Edinburgh magazine*, Nov 1832; 2, 8: 189–195.

Nichol, John P., 'On some recent discoveries in astronomy', *Tait's Edinburgh magazine*, 4.19, October 1833: 57–64.

Nichol, John P., 'The Irish Tithe Bill', *Tait's Edinburgh magazine*; Jul 1837; 4, 43: 402–404.

Nichol, John P., *Memorials from Ben Rhydding: concerning the place, its people, its cures* (London: Charles Gilpin, 1852).

Nichol, John P., *The Architecture of the Heavens* (London: John Parker, 1850).

Nichol, John P., *The Stellar Universe: views of its arrangements, motions, and evolutions* (Edinburgh: John Johnstone, 1848).

Nichol, John P., *Thoughts on some important points relating to the system of the world* (Edinburgh: William Tait, 1846).

Nichol, John P., *Views of Astronomy: seven lectures delivered before the Mercantile Library Association of New York in the months of January and February 1848* (New York: Greeley & McElrath, 1848).

Nichol, John P., *Views of the Architecture of the Heavens. In a series of letters to a lady* (Edinburgh: William Tait, 1837).

Nichol, John, *Thomas Carlyle* (London: Macmillan, 1892).

Noble, William, 'Richard A. Proctor', *The Observatory*, vol 11 (1888): 366–368.

Paine, Thomas, *The Age of Reason* (1794 repr. London: Freethought Publishing, 1880).

Paley, William and Edmund Paley, *The works of W. Paley, D.D. And an account of the life and writings of the author.* vol 1 (Longmans: London, 1938).

Paley, William, 'A Defence of the Considerations on the Propriety of Requiring a Subscription to the Articles of Faith', in Meadley, *Memoirs* (this pamphlet follows after page 216 of Meadley's book).

Paley, William, *Natural Theology or Evidences of the Existence and Attributes of the Deity collected from the appearances of nature* (Philadelphia PA: John Morgan 1802).

Paley, William, *Paley's Natural Theology*, vol II (London: Charles Knight, 1836).

Parsons, Charles (ed.), *The scientific papers of William Parsons, third Earl of Rosse 1800–1867* (Cambridge University Press, 2011).

Pascal, Blaise, *Pensées* (New York: E. P. Dutton, 1958).

Poe, Edgar A., 'The Murders in the Rue Morgue', *Tales* (London: Wiley & Putnam, 1846), 116–150.

Proctor, Richard A. and Arthur C. Ranyard, *Old and New Astronomy* (London: Longmans, Green, 1892), 15–30.

Proctor, Richard A., 'Astrology', in Richard A., *Myths and Marvels of Astronomy* (New York: G. P. Putnam's Sons, 1877), 1–52.

Proctor, Richard A., 'Death of Worlds', *Knowledge*, vol 3, 4 May 1883: 263.

Proctor, Richard A., 'Religion and the Unknown', *Knowledge*, vol 10, 1 July 1887: 197.

Proctor, Richard A., 'Science and Religion. By the Editor', *Knowledge*, vol 1, 4 November 1881: 3–4.

Proctor, Richard A., 'The Unknowable; or, the Religion of Science', *Knowledge*, vol 9, 1 December 1885: 37.

Proctor, Richard A., *Other suns than ours: a series of essays on suns – old, young and dead* (1887 repr. London: Longmans, Green, 1903).

Proctor, Richard A., *Other Worlds than Ours* (1870 repr. New York: D. Appleton, 1886).

Proctor, Richard A., *Our Place among Infinities* (London: Henry S. King, 1875).

Proctor, Richard A., *Saturn and its System* (London: Longman, Green, Longman, Roberts, & Green, 1865).

Proctor, Richard A., *Science Byways* (London: Smith, Elder, 1875).

Proctor, Richard A., *The Great Pyramid: observatory, tomb, and temple* (London: Chatto & Windus, 1883).

Ranyard, Arthur C., 'Richard Anthony Proctor', *Journal of the Royal Astronomical Society*, XLIX, 4 (Feb. 1889).

'Recent Astronomy', *British Quarterly Review* (6 Aug. 1847): 1–40.

Robinson, Thomas R., 'Account of the Three-feet Telescope' in Parsons (ed.), *scientific*, 14–19.

Robinson, Thomas R., 'Dr. Robinson's reply to Sir John Herschel', *The Athenaeum*, Oct 21, 1843; 834: 945–946.

Robinson, Thomas R., 'Dr. Robinson's reply to Sir John Herschel', *The Athenaeum*, Nov 18, 1843; 838: 1027–1028.

Robinson, Thomas R., 'On Lord Rosse's Telescope (Continued)', *Proceedings of the Royal Irish Academy (1836–1869)*, vol 3 (1844–1847): 114–133.

Robinson, Thomas R., 'On the Present Condition of the Earl of Rosse's Great Telescope', *Proceedings of the Royal Irish Academy (1836–1869)*, vol 4 (1847–1850): 119–128.

Romanes, Ethel, *Life and Letters of George John Romanes* (London: Longmans, Green, 1896).
Russell, Alexander, *Lord Kelvin: his life and work* (London: T. C. & E. C. Jack, 1912).
Schuster, Arthur, 'Memoir of the late Professor Balfour Stewart, LL.D., F.R.S. By Professor A. Schuster, Ph.D., F.R.S., F.R.A.S.,' in *Memoirs and Proceedings of the Manchester Literary & Philosophical Society*, Fourth Series, vol 1 (Manchester, 1888), 253–272.
Schuster, Arthur, *Biographical Fragments* (London: Macmillan, 1932).
Shelley, Percy B., 'The Necessity of Atheism' in Percy B. Shelley (foreword by Henry S. Salt), *Selected Prose Works of Shelley* (London: Watts, 1915), 1–15.
Shelley, Percy B., *Prometheus Unbound: a Lyrical Drama in four acts with other poems* (London: C and J Ollier, 1820).
Shelley, Percy B., *Queen Mab: a philosophical poem* (London, 1813), 171.
Smith, Joseph, *The Book of Mormon* (Palmyra NY: E. B. Grandin, 1830).
Spencer, Herbert, 'Consciousness under chloroform', *Popular Science Monthly*, 13, October 1878: 694–698.
Spencer, Herbert, 'Recent astronomy, and the nebular hypothesis', *Westminster Review*, July 1858; 70, 137: 185–225.
Spencer, Herbert, 'Recent astronomy, and the nebular hypothesis', *Illustrations of Universal Progress; a Series of Discussions* (New York: D. Appleton, 1865), 239–299.
Spencer, Herbert, 'The origin of animal-worship, etc.', *The Fortnightly Review*, May 1870 (vol 13 old series, 7 new series): 535–550.
Spencer, Herbert, *An Autobiography*, vol I (London: Williams and Norgate, 1904).
Spencer, Herbert, *An Autobiography*, vol II (London: Williams and Norgate, 1904).
Spencer, Herbert, *First Principles of a new system of Philosophy* (New York: D. Appleton, 1865).
Stair Douglas, Janet M., *The life and selections from the correspondence of William Whewell, D. D. late Master of Trinity College Cambridge* (London, 1881).
Stevenson, Robert L., 'Memoir' in Colvin and Ewing (eds.), *Papers*, vol 1, xi–cliv.
Stewart, Balfour and J. Norman Lockyer, 'The Sun as a Type of the Material Universe: Part II The Place of Life in a Universe of Energy', *Macmillan's Magazine*, 1 May 1868: 319–327.
Stewart, Balfour and Peter G. Tait, *The Unseen Universe: or Physical Speculations on a Future State* (London: Macmillan, 1875).
Stewart, Balfour and Peter G. Tait, *The Unseen Universe: or Physical Speculations on a Future State* (London: Macmillan, 1886).
Tait, Peter G., 'Does Humanity demand a New Revelation', *International Review* (November 1878).
Tait, Peter G., 'Dr. Balfour Stewart, F.R.S.', *Nature*, 29 December 1887: 202–203.
Tait, Peter G., 'Geological Time', *North British Review* 50 (1869): 406–439.

Tait, Peter G., *Lectures on some recent advances in physical science*, 2nd Edition (London: Macmillan, 1876).

Tayler, Charles B., *Sermons for all seasons, chiefly on the subject of Tractarian error* (London: Sampson Low, 1850).

Tennyson, Alfred Lord, 'Ode on the Death of the Duke of Wellington' in Hodder (ed.), *Tennyson*, 400–407.

Tennyson, Alfred Lord, 'In Memoriam A. H. H.' in Hodder (ed.), *Tennyson*, 305–389.

Tennyson, Alfred Lord, 'The Lotos-Eaters' in Hodder (ed.), *Tennyson*, 93–97.

Tennyson, Alfred Lord, 'The Princess' in Hodder (ed.), *Tennyson*, 227–303.

Tennyson, Alfred Lord, 'The Two Voices' in Hodder (ed.), *Tennyson*, 172–185.

Thompson, Silvanus P., *The Life of William Thomson: Baron Kelvin of Largs*, 2 volumes (London: Macmillan, 1910).

Thomson, William, 'On the age of the sun's heat', *Macmillan's Magazine*, 5 (1862): 388–393.

Thomson, William, 'On the Mechanical Energies of the Solar System.' *Edinburgh Royal Society Proceedings* [read 17 April 1854] vol III (1857): 241–244.

Thomson, William, 'On the Secular Cooling of the Earth', *Transactions of the Royal Society of Edinburgh*, [Read 28 April 1862] (Edinburgh: Robert Grant & Son, 1864), vol XXIII, 167–169.

Thomson, William, 'Physical considerations regarding the possible age of the Sun's heat', *Report of the Thirty-First meeting of the British Association for the Advancement of Science; held at Manchester in September 1861* (London: John Murray, 1862), 27–28.

Thomson, William, 'Physical Considerations regarding the Possible Age of the Sun's Heat', *The Athenaeum*, 1771, 28 September 1861: 412–413.

Thomson, William, *Mathematical and Physical Papers*, vol 1 (Cambridge University Press, 1882).

Thomson, William, *Mathematical and Physical Papers*, vol III (London: C. J. Clay and Sons, 1890), 295–311.

Todhunter, Isaac, *William Whewell, D. D. Master of Trinity College, Cambridge. An account of his writings with selections from his literary and scientific correspondence*, 2 volumes (London: Macmillan, 1876).

Tylor, Edward B., *Primitive Culture*, 2 volumes (London: John Murray, 1871).

Tyndall, John, 'Apology for the Belfast Address' in John Tyndall, *Fragments of Science: a series of detached essays, addresses and reviews*, vol II (New York: D. Appleton, 1892), 202–223.

Wells, Herbert G., 'Under the Knife', in *Tales of the Unexpected* (London: W. Collins Sons, 1924), 54–70.

Wells, Herbert G., *The Time Machine: An Invention* (London: Heinemann, 1895).

Whewell, William, 'Second Memoir on the Fundamental Antithesis of Philosophy', *Transactions of the Cambridge Philosophical Society*, vol VIII, part I (Cambridge: University Press, 1849), 614–662.

Whewell, William, *A Dialogue on the Plurality of Worlds* (London: John W. Parker and Son, 1853).

Whewell, William, *Astronomy and General Physics Considered with Reference to Natural Theology* (London: William Pickering, 1833).

Whewell, William, *Indications of the Creator* (London: John W. Parker, 1846).

Whewell, William, *Of the Plurality of Worlds: an Essay* (London: John W. Parker and Son, 1853).

Williams, Charles J. B., 'Affections of the Liver – Hepatitis', *The Boston Medical Surgery Journal*, 1845, 32: 149–155. https://doi.org/10.1056/NEJM184503260320801.

Willm, Joseph, *The Education of the People: a practical treatise on the realm of extending its sphere & improving its character. With a preliminary dissertation on some points connected with the present position of education in this country by J. P. Nichol, LL. D.* (Glasgow: William Lang, 1847).

Bibliography of Secondary Sources

7th Earl of Rosse, 'William Parsons, third Earl of Rosse', *Hermathena*, No. 107 (Autumn 1968): 5–13.

Alexander, Denis, *Rebuilding the Matrix: Science and Faith in the 21st Century* (Oxford: Lion 2002).

Alison, Countess of Rosse, 'Part 2: Family Life at Birr' in Mollan (ed.), *Parsons*, 55–70.

Aspasia Stephanou, Aspasia, 'Lovely Apparitions and Spiritualized Corpses: Consumption, Medical Discourse, and Edgar Allan Poe's Female Vampire', The Edgar Allan Poe Review, 14, 1 (Spring 2013), 36–54.

Baldwin, Melinda, 'The shifting ground of *Nature*: establishing an organ of scientific communication in Britain, 1800–1900', *History of Science* 1 (2012): 125–152.

Barton, Ruth, "John Tyndall, Pantheist: A Rereading of the Belfast Address', *Osiris*, 2nd Series, 3 (1987): 111–134.

Barton, Ruth, 'Scientific Authority and Scientific Controversy in *Nature*: North Britain against the X Club', in Louise Henson et al (eds.), *Culture and science in the nineteenth-century media* (Aldershot: Ashgate, 2004), 223–235.

Beer, Gillian, '"The Death of the Sun": Victorian Solar Physics and Solar Myth', in J. Barrie Bullen (ed.), *The Sun is God: Painting, Literature and Mythology in the Nineteenth Century* (Oxford: Clarendon Press, 1989).

Bennett, Jim A., *Church, State and astronomy in Ireland* (Belfast: Armagh Observatory, 1990).

Bernhardt, Boris C., and Tania Singer, 'The neural basis of empathy', *Annual Review of Neuroscience*, 35 (2012): 1–23.
Berridge, Victoria, *Demons: our changing attitudes to alcohol, tobacco, & drugs* (Oxford University Press, 2013).
Berridge, Victoria, *Opium and the People: Opiate Use and Drug Control Policy in Nineteenth and early Twentieth Century England* (London: Free Association Books, 1999).
Bisheff, Maurice, 'Paine, Spinoza, and the Radical Enlightenment', *Bulletin of Thomas Paine Friends*, 13, 2, Summer 2012: 7–8.
Bobis, Laurence and James Lequeux, 'Cassini, Rømer and the speed of light', *Journal of Astronomical History and Heritage*, 11 (2), 2008: 97–105.
Boddice, Rob and Mark Smith, *Emotion, Sense, Experience* (Cambridge University Press, 2020).
Boddice, Rob, *The History of Emotions* (Manchester University Press, 2018).
Brooke, John H., 'Darwin and Religion: Correcting the Caricatures', *Science & Education* (2010), 19: 391–405. https://doi.org/10.1007/s11191-009-9211-9.
Brooke, John H., 'Natural theology and the plurality of worlds: Observations on the Brewster-Whewell debate', *Annals of Science*, 1977, 34, 3: 221–286.
Brooke, John H., 'Science and the fortunes of Natural Theology: some historical perspectives', *Zygon*, 24, 1 (March 1989): 3–22.
Brooke, John H., *Science and religion: Some historical perspectives* (Cambridge University Press, 1991).
Brown, Callum G., *The Death of Christian Britain: Understanding Secularisation, 1800–2000* (London: Routledge, 2001).
Brown, Daniel, *The Poetry of Victorian Scientists: Style, Science and Nonsense* (Cambridge University Press, 2013).
Brown, Niamh, 'Devotional Cosmology: Poetry, Thermodynamics and Popular Astronomy, 1839–1889', PhD thesis submitted to the University of Glasgow (2016).
Brown, Stewart J., *Thomas Chalmers and the Godly Commonwealth in Scotland* (Oxford University Press, 1982).
Brush, Stephen G., *The Nebulous Earth: The Origin of the Solar System and the Core of the Earth from Laplace to Jeffreys* (Cambridge University Press, 1996).
Brush, Stephen G., 'The nebular hypothesis and the evolutionary world view', *History of Science*, 25 (1987): 245–278.
Buttmann, Günther, (tr. Bernard E. J. Pagel), *The Shadow of the Telescope: A Biography of John Herschel* (Guildford: Lutterworth, 1974).
Byrne, Katherine, *Tuberculosis and the Victorian literary imagination* (Cambridge University Press, 2011).
Campbell, James D., *Samuel Taylor Coleridge: A Narrative of the Events of his Life* (London: Macmillan, 1894).

Carlton, Eric J., *Patterns of Belief. Volume 2: Religions in Society* (London: George Allen & Unwin, 1973).
Chapman, Allan, 'William Parsons and the Irish nineteenth-century tradition of independent astronomical research' in Mollan (ed.), *Parsons*, 271–297.
Charbonneau, Paul, 'Stewart, Balfour' in T. Hockey (ed.), *Biographical Encyclopedia of Astronomers* (New York: Springer Science+Business Media, 2014), 2057–2059.
Checkland, Sydney and Olive Checkland, *Industry and Ethos: Scotland 1832–1914* (London: Edward Arnold, 1984).
Clarke, Martin L., *Paley: Evidences for the Man* (SPCK: London, 1974).
Crowe, Michael, J., *The extraterrestrial life debate 1750–1900: the idea of a plurality of worlds from Kant to Lowell* (Mineola NY: Dover Publications, 1999).
Daston, Lorraine and Peter Galison, *Objectivity* (New York: Zone Books, 2010).
Daston, Lorraine, 'The Moral Economy of Science', *Osiris* (01/1995) 10: 2–24.
Dawson, Gowan, *Darwin, Literature & Victorian Respectability* (Cambridge University Press, 2007).
Deng, Jiawen et al, 'The prevalence of depression, anxiety, and sleep disturbances in COVID-19 patients: a meta-analysis', *The Annals of the New York Academy of Sciences* 1486 (2020) 1–22. https://doi.org/10.1111/nyas.14506.
Desmond, Adrian and James R. Moore, *Darwin* (London: Michael Joseph, 1991).
Devine, Thomas M., *The Scottish nation: 1700–2000* (London: Allen Lane, 1999).
Dick, Steven J., 'Cosmic Evolution: History, Culture, and Human Destiny' in Steven J. Dick and Mark L. Lupisellsa (eds.), *Cosmos & Culture; Cultural Evolution in a Cosmic Context* (Washington DC: NASA, 2009).
Dick, Steven J., 'The Origins of the Extraterrestrial Life Debate and its Relation to the Scientific Revolution', *Journal of the History of Ideas*, 41, 1 (Jan–Mar., 1980): 3–27.
Dick, Steven J., *Plurality of Worlds: The Origins of the Extraterrestrial Life Debate from Democritus to Kant* (Cambridge University Press, 1982).
Dixon, Thomas, '"Emotion": The History of a Keyword in Crisis', *Emotion Review*, 4, 4 (October 2012): 338–344.
Dixon, Thomas, *From Passions to Emotions: The Creation of a Secular Psychological Category* (Cambridge University Press, 2003).
Dixon, Thomas, *Weeping Britannia: Portrait of a Nation in Tears* (Oxford University Press, 2015).
Dobson, Eleanor, 'Literature and Culture in the Golden Age of Egyptology', PhD Thesis submitted to the University of Birmingham (2016).
Dormandy, Thomas, *The White Death: a history of tuberculosis* (London: The Hambledon Press, 1999).
Dreyer, John L. E. and Herbert H. Turner (eds.), *History of the Royal Astronomical Society: Volume 1 1820–1920* (1923 repr. Oxford: Blackwell, 1987).

Drinko, Clayton D., *Theatrical Improvisation, Consciousness, and Cognition* (New York: Palgrave Macmillan, 2013).
Dror, Otniel E., Bettina Hitzer, Anja Laukötter and Pilar León-Sanz, 'Introduction to *History of Science and the Emotions*', *Osiris*, 31 (2016), 1–18.
Dror, Otniel, 'Historians in the Emotion Laboratory', *Emotion Review*, 12 (2020).
Earle L. and Seymour T. Porter, 'Samuel Taylor Coleridge and Opium', *Huntington Library Quarterly*, 17, 4 (Aug., 1954): 357–378.
Einstein, Albert, 'Maxwell's influence on the development of the concept of physical reality' in *James Clerk Maxwell: A Commemoration Volume 1831–1931* (Cambridge: The University Press, 1931), 66–73.
Elfenbein, Andrew, 'Samuel Taylor Coleridge, On the Constitution of the Church and State (1829)', *Victorian Review*, Spring 2009, 35, 1 (Spring 2009): 19–23.
Eliot, Thomas S., 'Introduction' in Pascal, *Pensées*, vii–xix.
Elliott, Paul, 'Erasmus Darwin, Herbert Spencer, and the Origins of the Evolutionary Worldview in British Provincial Scientific Culture, 1770–1850', *Isis*, 94, 1 (March 2003): 1–29.
Evans, Richard J., *The Pursuit of Power: Europe 1815–1914* (London: Penguin, 2016).
Feldman Barrett, Lisa, *How Emotions Are Made: The Secret Life of the Brain* (London: Pan Macmillan, 2018).
Feldman Barrett, Lisa, 'The theory of constructed emotion: an active inference account of interoception and categorization', *Social Cognitive and Affective Neuroscience*, 2017.
Feldman Barrett, Lisa, *Seven and a Half Lessons about the Brain* (London: Picador, 2021).
Fergusson, David A. S., 'Predestination: A Scottish Perspective', *Scottish Journal of Theology*, 1993, 46, 4: 457–478.
Foxcroft, Louise, *The Making of Addiction: The "Use and Abuse" of Opium in Nineteenth-Century Britain* (Aldershot: Ashgate, 2006).
Frye, Northrop, 'Defining an Age of Sensibility', *ELH (English Literary History)*, 23, 2 (June 1956): 144–152.
Fuller, Andrew, *The Gospel, its own witness: or the Holy Nature and Divine Harmony of the Christian Religion Contrasted with the Immorality and Absurdity of Deism* (New York: Isaac Collins, 1801).
Fyfe, Aileen, 'Publishing and the classics: Paley's *Natural theology* and the nineteenth-century scientific canon', *Studies in History and Philosophy of Science*, 33 (2002): 729–751.
Fyfe, Aileen, 'The Reception of William Paley's "Natural Theology" in the University of Cambridge', *The British Journal for the History of Science*, 30, 3, *Student Papers* (Sep., 1997): 321–335.
Gange, David, *Dialogues with the Dead: Egyptology in British Culture and Religion, 1822–1922* (Oxford University Press, 2013).

Gillham, Nicholas W., *A Life of Sir Francis Galton: from African Exploration to the Birth of Eugenics* (Oxford: Clarendon Press, 2001).

Gillispie, Charles G. (ed.), *Dictionary of Scientific Biography*, vol XIV (New York: Scribner, 1976).

Goldman, Martin, *The Demon in the Aether, The Story of James Clerk Maxwell* (Edinburgh: Paul Harris, 1983).

Gregory, Frederick, 'Intersections of Physical Science and Western Religion in the Nineteenth and Twentieth Centuries' in Mary J. Nye (ed.), *Cambridge History of Science*, vol 5, chapter 2 (Cambridge University Press, 2003): 36–53.

Gribbin, John, *The Reason Why: The Miracle of Life on Earth* (London: Allen Lane, 2011).

Griffiths, Roland R., William A. Richards, Matthew W. Johnson, Una McCann and Robert Jesse, 'Mystical-type experiences occasioned by psilocybin mediate the attribution of personal meaning and spiritual significance 14 months later', *Journal of Psychopharmacology*, 22, 6 (2008): 621–632. https://doi.org/10.1177/0269881108094300.

Guthke, Karl S., (tr. Helen Atkins), *The last frontier: imagining other worlds, from the Copernican revolution to modern science fiction* (Cornell University Press, 1990).

Hall, Catherine, *Civilising Subjects: metropole and colony in the English imagination 1830–1867* (Cambridge: Polity, 2002).

Harris, José, *Private Lives, Public Spirit: Britain 1870–1914* (Harmondsworth: Penguin, 1994).

Harrison, Peter, "'Science' and 'religion': constructing the boundaries", in Thomas Dixon, Geoffrey Cantor & Stephen Pumfrey (eds.), *Science and Religion: New Historical Perspectives* (Cambridge University Press, 2010), 23–49.

Hayward, Rhodri, *Resisting History: Religious Transcendence and the Invention of the Unconscious* (Manchester University Press, 2008).

Heffernan, William C., 'The Singularity of our Inhabited World: William Whewell and A. R. Wallace in Dissent', *Journal of the History of Ideas*, 39, 1 (Jan.–Mar., 1978): 81–100.

Henchman, Anna, '"The Globe we groan in": Astronomical Distance and Stellar Decay in In Memoriam', *Victorian Poetry*, 41, 1, Spring 2003: 29–45.

Hewitt, Rachael, *A Revolution of Feeling: the decade that shaped the modern mind* (London: Granta, 2017).

Hilton, Boyd, *The Age of Atonement: The Influence of Evangelicalism on Social and Economic Thought 1785–1865* (Oxford University Press, 2001).

Hoegaerts, Josephine, 'S-s-s-syncopation: Music, Modernity, and the Performance of Stammering (Ca. 1860–1930)', *Societies*, 2015, 5: 744–759. https://doi.org/10.3390/soc5040744.

Hoemann, Katie, Madeleine Devlin and Lisa Feldman Barrett, 'Comment: Emotions Are Abstract, Conceptual Categories That Are Learned by a Predicting Brain', *Emotion Review*, 12, 4 (October 2020): 253–255, quoted in Boddice and Smith, *Experience*, 34.

Hoffman, David C., '"The Creation We Behold": Thomas Paine's "The Age of Reason" and the Tradition of Physico-Theology', *Proceedings of the American Philosophical Society*, 157, 3 (September 2013): 281–303.

Hogan, Margaret, 'William Parsons' influence on the town and community of Birr' in Mollan (ed.), *Parsons*, 90–121.

Holmes, John, 'Lucretius at the Fin de Siècle: Science, Religion and Poetry', *English Literature in Transition*, 1880–1920, 51, 3 (2008): 266–280.

Hoppen, K. Theodore, *Ireland since 1800: Conflict and Conformity* (Harlow: Longman, 1999).

Hoskin, Michael A., 'Rosse, Robinson and the Resolution of the Nebulae', *Journal for the History of Astronomy*, 21, 4 (Nov., 1990): 331–344.

Hoskin, Michael A., 'William Herschel and the Construction of the Heavens', *Proceedings of the American Philosophical Society*, 133, 3 (Sep., 1989): 427–433.

Hoskin, Michael A., *Stellar Astronomy: Historical Studies* (Chalfont St. Giles: Science History Publications, 1982),

Huang, Hsiang-Fu, 'Commercial and Sublime: Popular Astronomy Lectures in Nineteenth Century Britain', PhD Thesis submitted to University College London (2015).

Hunter Brown, *William James on Radical Empiricism and Religion* (University of Toronto Press, 2000), 29.

Hutton, Ronald, *The Making of Oliver Cromwell* (New Haven CT: Yale University Press, 2021).

Huxley, Aldous, *The doors of perception; and, Heaven and hell* (London: Chatto and Windus, 1956).

Jaki, Stanley, *Planets and Planetarians: A History of Theories of the Origin of Planetary Systems* (Edinburgh: Scottish Academic Press, 1978), 123–124.

Jalland, Patricia, *Death in the Victorian Family* (Oxford University Press, 1996).

Jay, Mike, *Emperors of Dreams: Drugs in the Nineteenth Century* (Sawtry: Dedalus, 2000).

Kaplan, Fred, *Thomas Carlyle: a biography* (Cambridge University Press, 1983).

Kashatus III, William C., 'Thomas Paine: A Quaker Revolutionary', *Quaker History*, 73, 2 (Fall 1984): 38–61 at 41, 50–61.

Khademi, Hooman, Farin Kamangar, Paul Brennan and Reza Malekzadeh, 'Opioid Therapy and its Side Effects: A Review', *Arch Iran Med.* 2016; 19 (12): 870–876.

Koch, John T., *Celtic Culture: A Historical Encyclopedia* (Santa Barbara CA: ABC-Clio, 2006), 1671.

Landsman, Ned C., *From Colonials to Provincials, American Thought and Culture 1680–1760* (Cornell University Press, 1997).

Larsen, Timothy, *Crisis of doubt: honest faith in nineteenth-century England* (Oxford University Press, 2006).

Lawrence, Christopher J., 'The power and the glory: Humphry Davy and Romanticism' in Andrew Cunningham and Nicholas Jardine (eds.), *Romanticism and the sciences* (Cambridge University Press, 1990), 213–227.

Lawrence, Philip, 'Heaven and Earth – the relation of the nebular hypothesis to geology', in Yourgrau and Breck (eds.), *Cosmology*, 253–281.

Leedham-Green, Elisabeth, *A concise history of the University of Cambridge* (Cambridge University Press, 1996).

LeMahieu, Dan L., *The Mind of William Paley: A Philosopher and his Age* (University of Nebraska Press, 1976).

Leys, Ruth, *The Ascent of Affect* (University of Chicago Press, 2017).

Lightman, Bernard V, 'Constructing Victorian heavens: Agnes Clerke and the "New Astronomy"' in Barbara T. Gates and Ann B. Shteir (eds.), *Natural Eloquence: Women Reinscribe Science* (University of Wisconsin Press, 1997), 61–75.

Lightman, Bernard V., 'Victorian Sciences and Religions: Discordant Harmonies', *Osiris*, 2nd Series, 16, *Science in Theistic Contexts: Cognitive Dimensions* (2001): 343–366.

Lightman, Bernard V., '*Robert Elsmere* and the Agnostic Crises of Faith.', in Richard J. Helmstadter and Bernard Lightman (eds.), *Victorian Faith in Crisis* (London: Macmillan, 1990), 283–311.

Lovejoy, Arthur O., *The Great Chain of Being: A Study of the History of an Idea* (Harvard University Press, 2001).

Matus, Jill L., 'Trauma, Memory, and Railway Disaster: The Dickensian Connection', *Victorian Studies*, 43, 3 (Spring, 2001): 413–436.

McKeown, Thomas and R. G. Record, 'Reasons for the decline of mortality in England and Wales during the nineteenth century', *Population Studies* (1962) 16, 2: 94–122.

Meadows, Arthur J., *Science and controversy: a biography of Sir Norman Lockyer* (London: Macmillan, 1977).

Meyer, Basil, 'Till Death Do Us Part: The Consumptive Victorian Heroine in Popular Romantic Fiction', *The Journal of Popular Culture*, 37, 2, 2003: 287–308.

Mineka, Frances E. (ed.), *The Earlier Letters of John Stuart Mill 1812–1848 (Volume XII of the Collected Works of John Stuart Mill)* (University of Toronto Press, 1963).

Mollan, Charles (ed.), *William Parsons 3rd Earl of Rosse: Astronomy and the castle in nineteenth-century Ireland* (Manchester University Press, 2014).

Moore, James R. (ed.), *History, Humanity and Evolution: Essays for John C. Greene* (Cambridge University Press, 1989).

Moore, James R., 'Of Love and Death: why Darwin "gave up Christianity"' in Moore (ed.), *History*, 195–229.

Moore, James R., *The Post-Darwinian Controversies: a study of the Protestant struggle to come to terms with Darwin in Great Britain and North America, 1870–1900* (Cambridge University Press, 1979).

Morus, Iwan R., *When Physics became King* (University of Chicago Press, 2005).

Moscoso, Javier, 'Exquisite and Lingering Pains: Facing Cancer in Early Modern Europe' in Rob Boddice (ed.), *Pain and Emotion in Modern History* (London: Palgrave Macmillan, 2014), 16–35.

Murray, Alex, 'Vestiges of the Phoenix: De Quincey, Kant and the Heavens', *Victoriographies* 1.2 (2011): 243–260. https://doi.org/10.3366/vic.2011.0031.

Nahin, Paul J., 'Maxwell's grand unification', *IEEE Spectrum*, March 1992: 45.

Nall, Joshua, *News from Mars: Mass Media and the Forging of a New Astronomy, 1860–1910* (University of Pittsburgh Press, 2019).

Nasim, Omar W., 'Extending the Gaze: The Temporality of Astronomical Paperwork', *Science in Context*, 26, Special Issue 2 (June 2013): 247–277. https://doi.org/10.1017/S0269889713000057.

Nasim, Omar W., 'Observation, working images and procedure: the 'Great Spiral' in Lord Rosse's astronomical record books and beyond', *The British Journal for the History of Science*, 43, 3 (September 2010): 353–389. https://doi.org/10.1017/S0007087410000051.

Nasim, Omar W., 'Observing by hand', *Rheinsprung 11- Zeitschrift für Bildkritik*, 3 (May 2012): 66–74.

Nasim, Omar W., 'The 'Landmark' and 'Groundwork' of stars: John Herschel, photography and the drawing of nebulae', *Studies in History and Philosophy of Science*, 42 (2011): 67–84.

O'Connor, Ralph, *The Earth on Show: fossils and the poetics of popular science, 1802–1856* (University of Chicago Press, 2007).

Ogilvie, Marilyn B., 'Robert Chambers and the Nebular Hypothesis', *The British Journal for the History of Science*, 8, 3 (Nov., 1975): 214–232.

Oppenheim, Janet, *The other world: spiritualism and psychical research in England, 1850–1914* (Cambridge University Press, 1985).

Oswald, Margit E. and Stefan Grosjean, 'Confirmation Bias' in Rüdiger F. Pohl (ed.), *Cognitive Illusions: A Handbook on Fallacies and Biases in Thinking, Judgement and Memory* (Hove: Psychology Press, 2004).

Owen, Alex, *Place of Enchantment: British Occultism and the Culture of the Modern* (University of Chicago Press, 2004).

Parsons, Gerald, 'Biblical Criticism in Victorian Britain: From Controversy to Acceptance?' in Gerald Parsons (ed.), *Religion in Victorian Britain: Volume II Controversies* (Manchester University Press, 1988), 238–257.

Partridge, Christopher, *High Culture: Drugs, Mysticism, and the Pursuit of Transcendence in the Modern World* (Oxford University Press, 2018).

Plamper, Jan, 'The History of Emotions: An Interview with William Reddy, Barbara Rosenwein, and Peter Stearns', *History and Theory*, 49 (May 2010): 237–265.

Porter, Roy, *Enlightenment: Britain and the Creation of the Modern World* (London: Allen Lane, 2000).

Reddy, William, *The navigation of feeling: a framework for the history of emotions* (Cambridge University Press, 2001).

Reisenauer, Eric M., '"The battle of the standards": Great Pyramid metrology and British identity', 1850–1890, *The Historian*, 65 (2003): 931–978.

Richards, Robert J., *Darwin and the Emergence of Evolutionary Theories of Mind and Behavior* (University of Chicago Press, 1987).

Roper, Michael, 'Splitting in unsent letters: writing as a social practice and a psychological activity', *Social History*, 26, 3, October 2001: 318–339.

Rosenwein, Barbara H. and Riccardo Cristiani, *What is the History of Emotions?* (Cambridge: Polity, 2018).

Roxborogh, John, 'The Legacy of Thomas Chalmers', *International Bulletin of Missionary Research* (October 1999): 173–176.

Sadie Plant, Sadie, *Writing on Drugs* (New York: Picador, 2001).

Schaffer, Simon, 'Astronomers mark time: discipline and the personal equation', *Science in Context*, 2, 1 (March 1988): 115–145.

Schaffer, Simon, 'Celestial Eccentric', *Nature*, 333, 2 June 1988: 405–406.

Schaffer, Simon, 'Herschel in Bedlam: Natural History and Stellar Astronomy', *British Journal for the History of Science*, 13, 3 (1980): 211–239.

Schaffer, Simon, 'On astronomical drawing' in Caroline A Jones and Peter Galison (eds.), *Picturing science, producing art* (New York: Routledge, 1998), 441–474.

Schaffer, Simon, 'The History and Geography of the Intellectual World: Whewell's Politics of Language', in Menachem Fisch and Simon Schaffer (eds.), *William Whewell: A Composite Portrait* (Oxford: Clarendon Press, 1991), 201–232.

Schaffer, Simon, 'The Leviathan of Parsonstown: literary technology and scientific representation', in Timothy Lenoir (ed.), *Inscribing science: scientific texts and the materiality of communication* (Stanford: University Press, 1998), 182–222.

Schaffer, Simon, 'The nebular hypothesis and the science of progress' in Moore (ed.), *History*, 131–164.

Schechner, Sara J., *Time & Time Again: how science and culture shape the past, the present and future* (Cambridge MA: Collection of Historical Scientific Instruments, Harvard University, 2014), 14.

Schumpeter, Joseph A., *History of economic analysis* (New York: Oxford University Press, 1954).

Schweber, Silvan S., 'John Herschel and Charles Darwin: A Study in Parallel Lives', *Journal of the History of Biology*, 22, 1 (Spring, 1989): 1–71.

Secord, James A. 'Behind the veil: Robert Chambers and *Vestiges*' in Moore (ed.), *History*, 165–194.

Secord, James A., *Visions of Science: books and readers at the dawn of the Victorian age* (Oxford University Press, 2014).

Secord, James A., *Victorian Sensation: The Extraordinary Publication, Reception, and Secret Authorship of Vestiges of the Natural History of Creation* (University of Chicago Press, 2000).

Seth, Anil K., 'Interoceptive inference, emotion, and the embodied self.', *Trends in Cognitive Sciences*, 17, 11 (2013).

Shields, Andrew, 'The public life and political opinions of the 3rd Earl of Rosse' in Mollan (ed.), *Parsons*, 122–158.

Singer, Katherine, 'Stoned Shelley: Revolutionary Tactics and Women Under the Influence', Studies in Romanticism, Winter 2009, 48, 4: 687–707.

Smail, Daniel L., *On Deep History and the Brain* (University of California Press, 2008).

Smith, Crosbie and Matthew Norton Wise, *Energy and Empire: A biographical study of Lord Kelvin* (Cambridge University Press, 1989).

Smith, Crosbie, *The Science of Energy: a cultural history of energy physics in Victorian Britain* (London: Athlone, 1998).

Smith, Jonathan, 'De Quincey's Revisions to "The System of the Heavens"', *Victorian Periodicals Review*, 26, 4 (Winter, 1993): 203–212.

Snyder, Laura J., 'It's *All* Necessarily So: William Whewell on Scientific Truth', *Studies in History and Philosophy of Science*, 25, 5 (1994): 785–807.

Snyder, Laura J., 'Lord only of the ruffians and fiends'? William Whewell and the plurality of worlds debate', *Studies in History and Philosophy of Science*, 38 (2007): 584–592.

Snyder, Laura J., *Reforming Philosophy: a Victorian Debate on Science and Society* (Chicago: University of Chicago Press, 2006).

Spencer, Carole D., 'Holiness: The Quaker Way of Perfection', *Quaker History*, 93, 1 (Spring 2004) 123–147.

Stafford, William, *John Stuart Mill* (Basingstoke: Macmillan, 1998).

Stanley, Matthew, 'The Pointsman: Maxwell's Demon, Victorian Free Will, and The Boundaries of Science', *Journal of the History of Ideas*, 69, 3, July 2008: 467–491.

Steinicke, Wolfgang, 'Birr Castle observations of non-stellar objects and the development of nebular theories' in Mollan (ed.), *Parsons*, 210–270.

Steinicke, Wolfgang, 'The M51 mystery: Lord Rosse, Robinson, South and the discovery of spiral structure in 1845', *Journal of Astronomical History and Heritage*, 15, 1 (2012): 19–29.

Steinicke, Wolfgang, *Observing and Cataloguing Nebulae and Star Clusters: From Herschel to Dreyer's New General Catalogue* (Cambridge University Press, 2010), 104–109.

Stevenson, Catherine B., 'The Shade of Homer Exorcises the Ghost of De Quincey: Tennyson's "The Lotos-Eaters"', *Browning Institute Studies*, vol 10 (1982): 117–141.

Sturrock, John, *The Language of Autobiography: studies in the first person singular* (Cambridge University Press, 1993).

Tapp, Eddie, 'The Unwrapping of a Mummy' in A. Rosalie David (ed.), *Manchester Museum Mummy Project: Multidisciplinary Research on Ancient Egyptian Mummified Remains* (Manchester University Press, 1979), 83–94.

Theerman, Paul, 'James Clerk Maxwell and religion', *American Journal of Physics* 54, 312 (1986): 312–317.

Trevor-Roper, Hugh, 'The Invention of Tradition: The Highland Tradition of Scotland', in Eric Hobsbawm and Terence Ranger (eds.), *The Invention of Tradition* (Cambridge University Press, 2012), 15–42.

Turner, Frank M., 'Lucretius among the Victorians', *Victorian Studies*, 16, 3 (March 1973): 329–348.

Turner, Frank M., 'The late Victorian conflict of science and religion as an event in in nineteenth-century intellectual and cultural history', in Thomas Dixon, Geoffrey Cantor and Stephen Pumfrey (eds.), *Science and Religion*, 87–110.

Turner, Frank M., The Victorian Conflict between Science and Religion: A Professional Dimension, *Isis*, 69, 3 (Sep., 1978): 356–376.

Vance, Norman, *Bible and Novel: Narrative Authority and the Death of God* (Oxford University Press, 2013).

Weekes, Trevor C., 'Origin of the 3rd Earl's interest in astronomy' in Mollan (ed.), *Parsons*, 30–43.

Weekes, Trevor C., 'The Nineteenth-Century Spiral Nebula Whodunit', *Physics in Perspective*, 12 (2010): 146–162.

Wheeler, Michael, *Heaven, Hell, & the Victorians* (Cambridge University Press, 1994).

Wheeler-Barclay, Marjorie, *Science of Religion in Britain, 1860–1915* (University of Virginia Press, 2010).

Whitrow, Gerald J., 'The role of time in cosmology', in Wolfgang Yourgrau & Allen D. Breck (eds.), *Cosmology, History and Theology* (New York: Plenum Press, 1977), 159–177.

Wilde, Norman, 'Scepticism and Faith in the Philosophy of Pascal', *The Harvard Theological Review*, 9, 1 (Jan., 1916): 56–83.

Winston, Mahabir, 'History of the Psychedelic Drug Experience', *Canadian Psychiatric Association Journal*, 13, 2 (1968): 189–190.

Yeo, Richard, 'William Whewell, natural theology and the philosophy of science in mid nineteenth century Britain', *Annals of Science*, 1979, 36, 5: 493–516.

Young, Robert J. C., *Colonial desire: hybridity in theory, culture and race* (London: Routledge, 1995).

Yule, John D., 'The impact of science on British religious thought in the second quarter of the nineteenth century', PhD thesis submitted to Cambridge University (1976).

Index[1]

[1] Note: Page numbers followed by 'n' refer to notes.

H. Carlton, *Cosmology and the Scientific Self in the Nineteenth Century*, https://doi.org/10.1007/978-3-031-05280-4

O

P

U

V

Kolibri 360 GmbH
Ferdinand-Jühlke-Straße 7
99095 Erfurt, Deutschland
produktsicherheit@kolibri360.de